Solving the Mystery of the Carolina Bays

Antonio Zamora

This work analyzes the geologic and geometrical characteristics of the Carolina Bays to derive their mechanism of formation. The prototypical shapes of the Carolina Bays are conic sections, and this implies that they must have originated from oblique conical cavities that were remodeled by geologic processes.

ISBN: 978-0-9836523-9-7
Paperback Edition 3

CONTENTS

INTRODUCTION

Have any humans ever been killed by a meteorite impact? Astronomers tell us that the collision of the Earth with a large asteroid or comet could cause a major extinction event, such as the one that killed the dinosaurs 65 million years ago. The explosion of a near-Earth asteroid over the city of Chelyabinsk, Russia on February 15, 2013 injured over 1400 people and served as a reminder of how vulnerable the Earth is to cosmic impacts. Some scientists have suggested that the explosion of a large extraterrestrial object over the ice sheet that covered North America during the ice ages 12,900 years ago caused the disappearance of the Clovis culture in North America, and that the Carolina Bays were produced as a result of this explosion (Firestone 2007, 2009). Understanding the origin of the Carolina Bays would clarify the history of North America, and would help to determine if the Clovis people and the megafauna that inhabited what is now the United States died as a result of an extraterrestrial impact. The Carolina Bays may be the key for solving this mystery.

The consensus among geologists is that the Carolina Bays were created over many millennia by the action of water and wind currents. The formation of the bays by a single event, such as an impact, has been ruled out based on the wide variability in the dates of the terrain on which the bays are found and on the fact that the bays do not contain meteorite fragments or the type of evidence that is characteristic of hypervelocity extraterrestrial impacts.

The evidence against impacts is considered so strong in the scientific community, that the suggestion that the Carolina Bays were created by impacts immediately raises great consternation and the type of hostile skepticism that is usually reserved for conspiracy theories and pseudoscientific arguments. Without being able to consider impacts, only geological explanations are accepted. However, the current geological explanations about the origin of the Carolina Bays leave many questions unanswered, and this provides the opportunity to develop new scientific hypotheses that are more comprehensive and fit the physical data better.

There are two features of the Carolina Bays that cast doubt on the current geological explanations. The first one is the radial alignment of the bays. Lines drawn along the major axes of the bays intersect at a common point in the Great Lakes region. A well-known characteristic of impacts is the radial distribution of ejecta emplaced along ballistic trajectories originating from the point of impact. One current geological hypothesis proposes that the alignment of the bays is due to action of the prevailing winds and wave erosion during more than one hundred thousand years of bay formation (Brooks et al. 2010). This rather simple explanation has been generally accepted, although the direction and strength of the wind during this extensive time period is not known and can only be guessed from geological records. The current explanations are also deficient because they consider only the bays along the Atlantic seaboard, but this does not provide a comprehensive hypothesis because Nebraska also has bays, called rainwater basins, and they have similar physical characteristics as the bays on the East Coast (Zanner 2001). The Nebraska bays are also aligned with their major axes pointing toward the Great Lakes region.

The bays in the East Coast are aligned in a northwest to southeast direction while the Nebraska bays are aligned in a northeast to southwest direction, almost perpendicular to those on the East Coast. The Nebraska bays occur at altitudes of 400 to 900 meters above sea level and far away from the sea. If the wind was responsible for the shape of the bays, does that mean that the wind in Nebraska was blowing in a direction perpendicular to the wind in the

East Coast for many thousands of years? This is not very likely and there is no evidence to support such assumption. The hypothesis that the bays were formed by wind currents and wave erosion also fails to explain why the bays with the older dates, which were exposed to erosive forces for a longer time, have exactly the same geometrical shape as the bays with younger dates. The action of wind or water should have had a greater effect on older bays than on new bays. One would expect that 60,000 or 80,000 years of wind and water erosion would have modified old bays substantially to give them different characteristics and a different shape from the younger bays.

The geometrical shape of the Carolina Bays is another strong reason for doubting the current geological explanations. The bays, whether in Nebraska or the East Coast have a remarkably consistent elliptical shape regardless of their size. Both, small and large bays have similar elliptical shapes with width-to-length ratios that average approximately 0.58. These width-to-length ratios are maintained even for bays that overlap and for small bays contained within larger ones. Most of the bays have elliptical shapes with similar aspect ratios. The bays are not just oval-shaped; they are elliptical in the mathematical sense and can be considered conic sections. This mathematical oddity of the Carolina Bays is as significant as the spherical shape of planets, but it has not been studied seriously. The current geological hypotheses do not have an explanation for this geometrical characteristic and they cannot explain how such precise shapes could have been made by wind and water processes. The geological hypotheses do not have physical or mathematical models to explain how water and wind can carve ellipses with specific width-to-length ratios and maintain their shapes for thousands of years in the vast area of the eastern half of the United States.

An adequate hypothesis about the formation of the

Carolina Bays should be able to explain the radial alignment of the bays and the reason why the bays have elliptical shapes with specific aspect ratios. In addition, if the bays were indeed created by ejecta from an impact, the hypothesis must be able to explain why the bays do not have the accepted markers for impacts and why they have such a wide range of dates. Solving the mystery of the Carolina Bays requires a multidisciplinary approach involving astronomy, physics and geology. The following discussion presents information about the Carolina Bays within a framework that satisfies all these criteria and is supported by an experimental physical model. Most of the images of the bays in this book have a scale bar and the latitude and longitude of the center of the image to enable readers to find the features easily with Google Earth.

***

WHAT ARE THE CAROLINA BAYS?

Many different types of lakes or marshy depressions have been erroneously classified as Carolina Bays. The Carolina Bays have also been characterized as similar to thermokarst or thaw lakes that are circular or elliptical in shape and are sometimes aligned with the prevailing wind (Melosh 2011). However, the Carolina Bays along the southern Atlantic coast occur on ground that never had permafrost even during the ice ages, so that mode of origin cannot be correct for the bays. The following definition of Carolina Bays is rather restrictive, but it is the only definition consistent with the physical characteristics of the bays for which a comprehensive mechanism of formation can be developed.

The Carolina Bays are shallow elliptical depressions with raised rims on unconsolidated ground whose major axis is oriented toward the Great Lakes region. The prototypical Carolina Bays are elliptical in the mathematical sense and they have an average width-to-length ratio of approximately 0.58. Carolina Bays only occur within a radius of 1500 kilometers from the Great Lakes.

This definition essentially constrains the occurrence of Carolina Bays to the eastern half of the United States – from Nebraska to the East Coast. The bays were formed during the Pleistocene Epoch and no bays formed north of the Great Lakes because at this time Canada was covered with a layer of ice hundreds of meters thick. The distance limit of 1500 kilometers is based on the empirical fact that no bays meeting all the other criteria have been found beyond this range.

Discovery of the Carolina Bays.

The Carolina Bays were known to the early settlers of the United States as marshy areas with sandy rims. A comprehensive bibliography of the Carolina Bays (Ross, 1987) shows that from the middle of 19th century to the early 20th century the bays were characterized as spring basins, sand bar dams of drowned valleys or as depressions dammed by giant sand ripples. The bays are very large and their rims are usually only one meter higher than the centers so they are virtually undetectable from the ground. For this reason, many of the bays remained undiscovered at this time.

All this changed with the advent of aviation. For the first time, it became possible to see that the Carolina Bays were elliptical depressions that were all aligned in approximately the same direction. **Figure 1** is a satellite image from *Google Earth* of an area just north of White Oak, North Carolina. This is the type of landscape that was photographed by the first aerial mapping of the United States and which launched new studies about the bays. The faint outlines of the bays can be seen against a mosaic of farmland with plowed fields.

One of the first impact hypotheses for the formation of the Carolina Bays was made by Melton and Schriever from the University of Oklahoma in 1933. They suggested that a meteorite shower or a colliding comet coming from the northwest could have created the bays with their peculiar alignment. Their proposal was dismissed because the Carolina Bays have axial directions that are not parallel as would have been expected for impacts by extraterrestrial projectiles (Johnson 1942). Another objection was the gigantic size of the meteorites required to make the large bays and the lack of meteorite fragments in the coastal area where the Carolina Bays are found. This was a time when geologists were beginning to characterize extraterrestrial impacts.

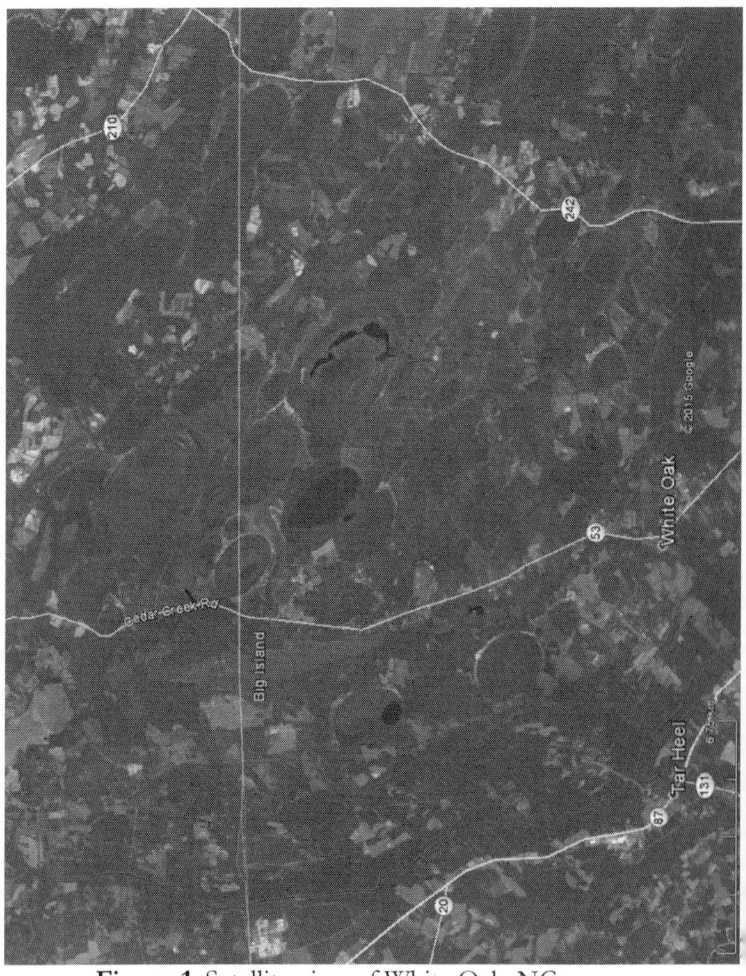

Figure 1. Satellite view of White Oak, NC
(Lat. 34.8258, Lon. -78.6974)

Surface structures created by impacts only gained general acceptance around 1960, when geologist Eugene M. Shoemaker presented criteria for establishing that Meteor Crater in Arizona was the result of an extraterrestrial impact and not the caldera of an extinct volcano. Another impact hypothesis that did not receive much attention suggested that shockwaves from cometary airbursts created the depressions (Eyton and Parkhurst, 1975).

Geologists continued proposing theories to explain the origin of the Carolina Bays that included the formation of lakes in sand elongated in the direction of maximum wind velocity, solution depressions with wind-drift sand forming the rims, basins scoured out by confined gyroscopic eddies, wind-driven (eolian) blowouts, and even fish nests made by giant schools of fish waving their fins in unison over submarine artesian springs. The origin of the Carolina Bays has remained a controversial subject and none of the numerous proposed theories has gained wide acceptance, although the combination of wind and water action is favored.

Many Carolina Bays have been destroyed by erosion, urbanization and farming, but most of their main features have been preserved. Here is a summary of the main morphologic characteristics of the Carolina Bays (Eyton and Parkhurst 1975):

1. The Carolina Bays are ellipses although some lack bilateral symmetry along either the major or the minor axis. The southeast portion of many bays is more pointed than the northwest end and the northeast side bulges slightly more than the southwest side. Known major axis dimensions vary from approximately 60 meters to 11 kilometers.
2. The Carolina Bays display a northwest-southeast orientation. Deviations from this orientation appear

to be systematic by latitude (Prouty, 1952).

3. The bays are shallow depressions below the general topographic surface with a maximum depth of about 15 meters. Large bays tend to be deeper than small bays, but the deepest portion of any bay is offset to the southeast from the bay center.
4. Many bays have elevated sandy rims with maximum development to the southeast. Rim heights vary from 0 to 7 meters.
5. Carolina Bays frequently overlap other bays without destroying the morphology of either depression. One or more small bays can be completely contained in a larger bay.
6. The stratigraphy beneath the bays is not distorted (Preston and Brown, 1964; Thom, 1970).
7. Bays occur only in unconsolidated sediments. Bays in South Carolina are found on relict marine barrier beaches associated with Pleistocene sea level fluctuations, in dune fields, on stream terraces and sandy portions of backbarrier flats (Thom, 1970). No bays occur on modern river flood plains and beaches.
8. Carolina Bays appear to be equally preserved on terraces of different ages and formational processes.
9. Bays are either filled or partly filled with silt of organic and inorganic origin. Ghosts of semi-obliterated Carolina Bays appear to represent former bays that were filled by terrestrial sediments and organic materials.

The following graph (**Figure 2**) displays the frequency of the various sizes of bays from a survey of approximately 500 bays. The graph shows that bays with major axes of approximately 250 meters are the most common. Approximately 65% of the Carolina bays have major axes smaller than 460 meters.

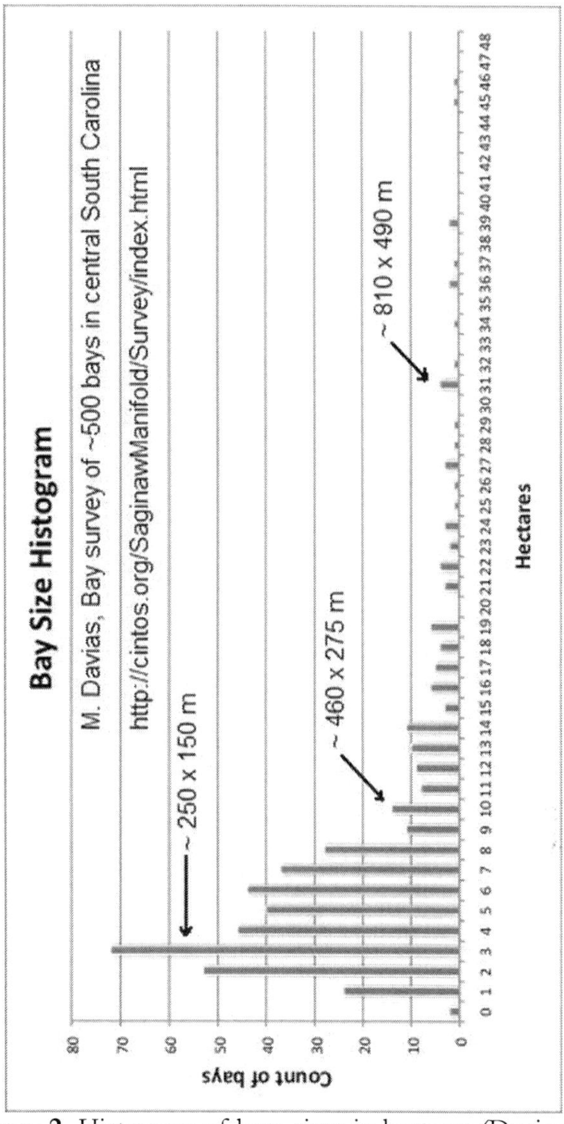

Figure 2. Histogram of bays sizes in hectares (Davias [1]) with approximate dimensions of major and minor axes

The graph helps to put things in perspective. Bays that are larger than one kilometer are very prominent in the images, but they comprise only a small percentage of the total number of bays.

Dates of the Carolina Bays

According to Brooks (2010), the Carolina Bays formed during the Pleistocene as shallow lakes and developed their northwest to southeast orientation by the action of southwesterly winds. The depressions were expanded and oriented by wave erosion, resulting in bay elongation perpendicular to the wind direction. The bays evolved as a result of processes active episodically for about 140 thousand years (ka). Based on 45 Optically Stimulated Luminescence (OSL) dates, active shorelines and associated eolian deposition occurred during Marine Isotope Stage (MIS) 2 to late MIS 3 (~12 to 50 ka), MIS 4 to very late MIS 5 (60-80 ka), and late MIS 6 (120-140 ka). The Marine Isotope Stages are alternating warm and cool periods in the Earth's paleoclimate. The temperature is estimated based on the ratio between oxygen-18 and oxygen-16 isotopes in deep-sea core samples.

The following image (**Figure 3**) shows a USGS topographic map of Big Bay, a Carolina Bay just north of Pinewood, SC. The center of the bay is located at approximately Lat. 33.78645, Lon. -80.46785, and the bay measures 4,920 by 3,100 meters. This gives it a width-to-length ratio of 0.63, which is close to the average for Carolina Bays.

Brooks states that Big Bay formed from wind-driven sand sheets from the Wateree River, which is 10 kilometers to the west. The sand moved across Big Bay about 74,000 years ago and was resurfaced subsequently 33,000 to 29,000 years ago. The innermost sand rim at Big Bay was remodeled as recently as 2,200 years before the present.

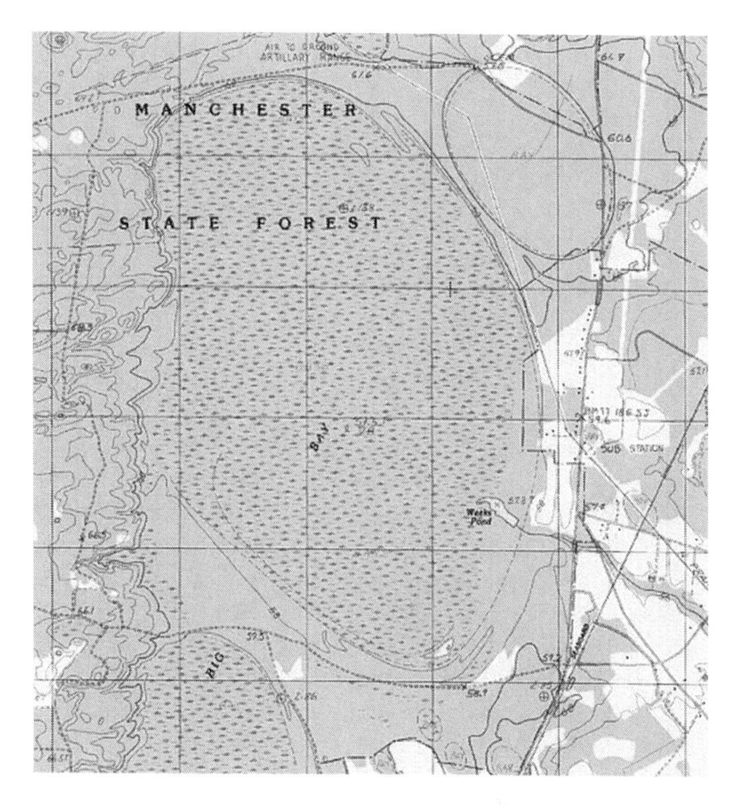

Figure 3. Big Bay, north of Pinewood, South Carolina (Lat. 33.78645, Lon. -80.46785)

Radial Alignment of the Carolina Bays

Early researchers noticed that the axial orientation of the Carolina Bays varied depending on their location along the East Coast of the United States. Prof. Douglas Johnson (1942) examined the orientation of the major axes of 381 bays in North Carolina, South Carolina, and Georgia. He found a marked contrast in the axial directions in the northern and the southern parts of this area. In the northern part of the area, the average axial directions trended toward the southeast, and in the southern part of the area, the orientation trended strongly toward the south. Johnson attributed this change in orientation as validation of his hypothesis involving artesian spring migration updip.

Eyton and Parkhurst (1975) measured the azimuths of 358 bays from Georgia to Virginia. Their analysis confirmed that the axial orientation of the bays varies systematically by latitude. A chart of the average orientations of the bays on a flat-earth map produced a radial alignment with an apparent focus in either southern Ohio or Indiana. Since not all the lines seemed to originate from a common point, the authors proposed that the majority of the bays with similar orientations had been created by fragments from a disintegrating comet, and that the bays with divergent orientation resulted from explosions and shock waves generated by further fragmentation of the remaining nucleus.

These orientation studies were done on a relatively small number of bays on the Atlantic seaboard before Zanner and Kuzila (2001) had reported the similarity of the Nebraska Rainwater Basins to the Carolina Bays. The Nebraska bays have an axial orientation almost perpendicular to the bays on the East Coast and they provide additional data points for triangulation of the focal point of the bays.

The emergence of satellite imagery from the Google Earth Geographic Information Systems (GIS) augmented with high resolution elevation data using laser and radar technology (LiDAR) made it possible to get more accurate triangulation of the convergence point of the axial directions of the Carolina Bays (Davias and Gilbride 2010). Taking into consideration great circle ejecta trajectories and adjusting for the rotation of the Earth, the epicenter or intersection of the axial orientations was found in Saginaw Bay in Michigan.

Davias and Harris (2015) showed that the orientation of any given bay can be predicted using simple trigonometry. The aviation bearing from a bay at point **A** toward the epicenter **B** (Lat. 43.68, Lon. -83.82) located in Saginaw Bay can be computed with the following formula which applies a Coriolis adjustment to compensate for the rotation of the Earth.

Great Circle Bearing at **A** =
180 +(ATAN2(COS(lat**A**)*SIN(lat**B**)
-SIN(lat**A**)*COS(lat**B**)*COS(lon**B**-lon**A**),
SIN(lon**B**-lon**A**)*COS(lat**B**)) * 180/PI())

The equation works equally well for the Rainwater Basins in Nebraska. The formula's only input variables are the latitude and longitude of the bay and the epicenter in Saginaw Bay. A graph of the measured orientations of 45,000 bays clusters neatly along the line calculated by the formula.

The fact that one equation can predict the average azimuthal orientation of such a large number of bays provides strong support for a common origin of the Carolina Bays and Nebraska Rainwater Basins as secondary impact structures produced by material radiating from an epicenter in Michigan. The material could have consisted of chunks of glacier ice ejected by a meteorite impact on the Laurentide ice sheet.

None of the hypotheses proposing that the bays were formed by terrestrial wind and water mechanisms have any formulation that can predict the axial orientations of the bays. In fact, there are conflicting opinions. Some authors state that the bays are aligned with the prevailing wind (Melosh 2011, Raisz 1934), while others claim that the bays are elongated perpendicular to the wind direction (Brooks 2010).

* * *

FIELD TRIP TO THE CAROLINA BAYS

Myrtle Beach has some of the most prominent Carolina Bays in the eastern shore of the United States. The bays in this region are also some of the most accessible because many roads and neighborhoods are close to well-preserved bays. The following description documents a two-day field trip to the Carolina Bays starting on July 4, 2014.

Figure 4 shows a rectangular area parallel to the Atlantic Ocean bounded by highway US-501 (Edward E. Burroughs Highway) on the southwest, SC-90 on the northwest, SC-22 (Conway Bypass) on the northeast and SC-31 (Carolina Bays Parkway) on the southeast. Almost any point in this area can be reached by car from Myrtle Beach within 30 minutes.

The map from Google shows the major roads in the area, including North Ocean Boulevard where many oceanfront hotels are located.

In preparation for the trip, I loaded my Garmin GPS unit with the coordinates of some of the roads that I wanted to explore. I was particularly interested in a branch of Old State Highway 90 that goes along the rim of Watts Bay (**Figure 5**). The Google map indicated that the road dead-ended at the terminal end of the bay. I thought that this might be a good spot to take a soil sample. I also consulted the U.S. Geological Survey map of the area prior to the trip (**Figure 6**). The names of the larger bays are recorded in the maps of the U.S. Geological Survey.

Figure 4. Satellite view of the Myrtle Beach area showing the major roads

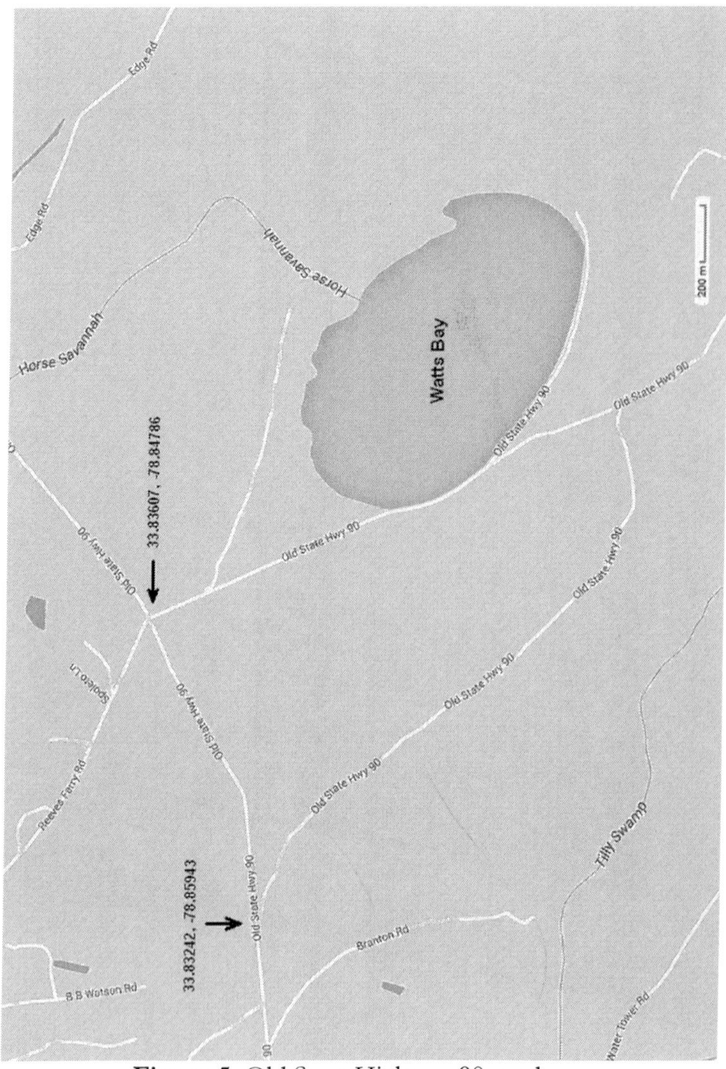

Figure 5. Old State Highway 90 roads by Watts Bay

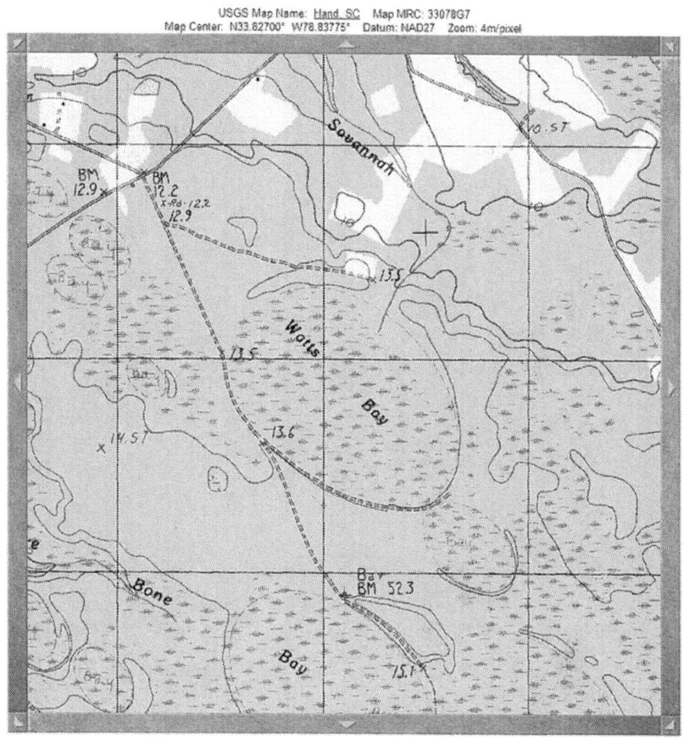

Figure 6. USGS topographic map of Watts Bay

July 4, 2014

Traveling along the Conway Bypass road (SC-22), it is easy to miss Old State Highway 90 because there are no signs marking it.

Figure 7. Old State Highway 90, just off SC-22

It was only through GPS navigation that I knew that the dirt road was Old State Highway 90. Soon after I got on this road, the GPS informed me that I was driving "off-road", although it still displayed the map. Hurricane Arthur had passed near the coast two days before. The road was still quite wet and the car skidded in some spots as it tried to get traction on the loose sandy soil.

Figure 8. Intersection of Reeves Ferry Road and Old State Highway 90

I reached my turning point by Reeves Ferry Road (**Figure 8**) and discovered that the road that I planned to use was closed.

Figure 9. Old State Highway 90 branch

The Old State Highway 90 branch road that I wanted to use to go around the bay was closed with a cable (**Figure 9**), and a "No Trespassing" sign was posted. I continued west on Old State Highway 90 to a second branch near Branton Road.

Figure 10. Old State Highway 90 branch near Branton Rd.

The Old State Highway 90 branch near Branton road (**Figure 10**) was also closed with a sign that said "Private Property - no trespassing". Obviously, it was not going to be possible to use these old highway branches to explore Watts Bay.

Figure 11. Power lines along
Old State Highway 90 branch near Branton Rd.

Fortunately, there was an electric power line easement crossing Old State Highway 90 branch near Branton Road (**Figure 11**). This was a place where it was possible to get a sample of the soil that is typical of where the Carolina Bays are found.

Figure 12. Soil sample from the Carolina Bays

Figure 12 shows me holding a shovel of soil typical of the Carolina Bays. The soil is moist and sandy; it does not have any cohesion.

Figure 13. Hole from which the sample was taken

The soil in this area is almost like the sand found on the beach except for a thin surface layer with coloration from organic matter. The cross-section of the hole from which a sample was taken shows that the top 10 centimeters have a slightly darker color (**Figure 13**). The sand at a deeper level is similar to the sand in the beach by the Atlantic Ocean.

Although the soil in this tract of land has been cleared of vegetation for the easement of the power lines, the current soil is probably representative of the soil in the bays because the removal of the vegetation disturbed only the surface of the soil, but no extraneous soil was brought as filler.

Figure 14. Packed sand circles on the road

After I got home and I examined my photographs of the visit to Myrtle Beach, I noticed something that had escaped my attention when I was in the field. A section of the dirt road on which I drove had some circles of packed sand. Judging from the tire tracks, the circles had a harder surfacc than the surrounding area. The circles were probably created when subsurface water from the rains brought by hurricane Arthur flowed through the ground creating a quicksand pool that became a smooth sandy circle after the water level dropped. This may be another phenomenon besides viscous relaxation that could have leveled out the Carolina Bays.

July 5, 2014

After having been thwarted from traveling along the edge of Watts Bay, I went back to Google maps and chose a new target bay near a residential neighborhood close to US-501.

Figure 15. Cottonpatch Bay
(Lat. 33.781279, Lon. -78.944193)

Cottonpatch Bay (**Figure 15**), centered at Lat. 33.781279 and Lon. -78.944193 has a major axis of 1920 meters (6299 ft.) and a minor axis of 1175 meters (3855 ft.). The ratio of the minor to the major axis of 0.61 is close to the average for Carolina Bays. I chose Cottonpatch Bay because it had a road along its perimeter.

The South Prong Steritt Swamp is located just to the west of the bay. The fact that a swamp is next to the bay provides an indication of how close the water table is to the surface. The northwest portion of Cottonpatch Bay has been cleared of vegetation and a grid of channels has been dug to drain the land. Eventually, this portion of the ellipse may be filled with rows of houses.

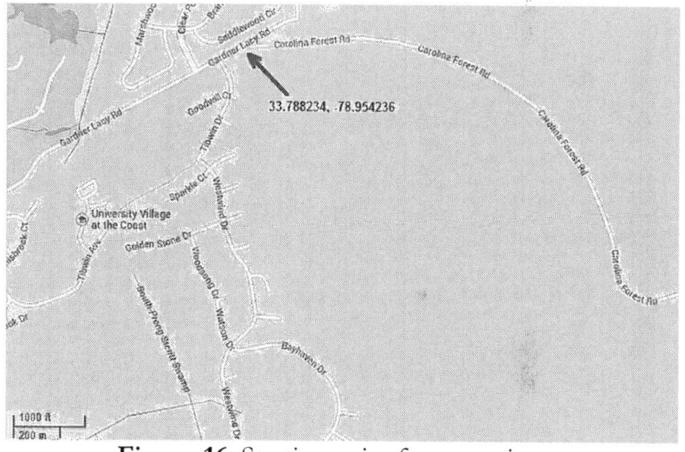

Figure 16. Starting point for excursion along the rim of Cottonpatch Bay.

Setting the GPS to 1599 Gardner Lacy Road helped me to navigate to the intersection of Gardner Lacy Road and Carolina Forest Road that goes along the border of the bay (**Figure 16**). The USGS topographic map of Cottonpatch Bay is shown in **Figure 17**.

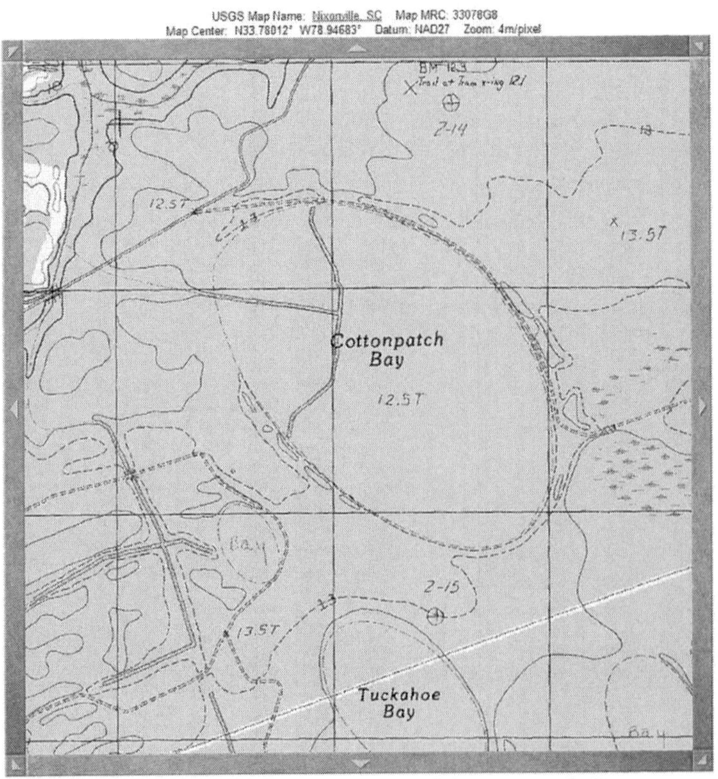

Figure 17. USGS topographic map of Cottonpatch Bay

Figure 18. Intersection of Gardner Lacy Road and Carolina Forest Road

Carolina Forest Road starts out as a gravel road that eventually becomes just a sandy road (**Figure 18**). There is no sign of any kind marking the road. It can only be identified from GPS coordinates.

Some distance from the start of the road, there was a gate, but it was open (**Figure 19**). The gate was curved, as if at some time in the past a vehicle had crashed into it trying to get out. This road had no signs forbidding passage, so I drove along the single-lane road toward the bend shown in the map.

Figure 19. Open gate without warning signs

The deep ditch along the road keeps the road from flooding during rainy weather.

Figure 20. Bay to the right, rim to the left

Figure 20 is typical of what a Carolina Bay looks like. The bay, to the right of the road, has a thick growth of vegetation consisting mostly of shrubs. Trees usually do not survive within the bays due to the wet soil conditions. The root systems of the dense vegetation have probably contributed to the preservation of the structure of the bays. The sandy rim, to the left of the road, has mostly grass and trees.

Figure 21. Bayberry bushes

Bayberry bushes of the genus *Myrica* densely cover the Carolina Bays (**Figure 21**). Bayberry leaves and the waxy gray berries, which are the size of a peppercorn, have an aromatic scent that is used in candles, soaps and perfumery.

Figure 22. Sandy rims of the Carolina Bays

The sandy rims of the bays are sparsely covered with grasses and trees (**Figure 22**). The poor soil does not support much vegetation.

Cottonpatch Bay is under commercial development. There was a bulldozer and a tank car parked at the bend of Carolina Forest Road. Although the road continued deeper into the woods, my car was not getting good traction on the soft ground. Carefully, I turned the car around and went back to my hotel to enjoy some time at the beach. A 4-wheel drive vehicle would have been more appropriate for this excursion.

My field trip to the Carolina Bays was not as simple as I had imagined because many of the bays are in private property. My trip also revealed the high rate at which farming and residential construction are destroying the bays, and this will make it more difficult to conduct meaningful research on the structure of the bays in the future.

* * *

THE CAROLINA BAYS ARE TRUE ELLIPSES

In the early 1960s, the invention of LiDAR combined laser-focused imaging with distances calculated by radar. This new technology vastly improved the detection of Carolina Bays by emphasizing small differences in the elevation of surface features. When used on the east coast of the United States, LiDAR discovered a great number of bays along the Atlantic seaboard in Delaware, Maryland, New Jersey, North Carolina, South Carolina, Virginia, Georgia, and north central Florida. It is estimated that there are approximately 500,000 Carolina Bays. In some areas, the bays are so dense that all the ground is completely covered by them, except where streams and erosion have washed them away.

Figure 23 shows Carolina Bays southwest of Fayetteville, NC with many of the features that distinguish them, such as the regular elliptical shape and the raised rims with the thickening in the southeast border of the bay. Some of the bays overlap while retaining their characteristic width-to-length ratio.

Just as aerial photography and LiDAR enhanced the study of the Carolina Bays, the introduction of *Google Earth* in 2001 brought significant advances. The satellite images from Google made it possible for anybody with access to the Internet to explore the surface of the Earth and measure natural and man-made structures. With these resources, Michael Davias established a web site for viewing Carolina Bay LiDAR images (Davias [1]).

The geological similarity between the Nebraska Rainwater Basins and the Carolina Bays was first described by Zanner and Kuzila in 2001. Measurement of the width and length of the Carolina Bays and the Nebraska Rainwater Basins can be used to establish that the similarity also encompasses their elliptical geometrical characteristics.

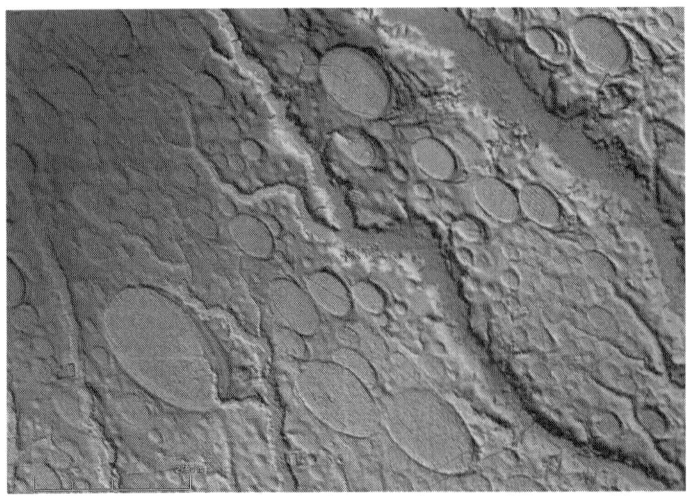

Figure 23. Carolina Bays 37 km southwest of Fayetteville, NC viewed with LiDAR (Lat. 34.850, Lon. -79.205)

The width-to-length ratios of the Carolina Bays average 0.58 and are very consistent for bays of different sizes. The bays in Nebraska are indistinguishable from the bays in the East Coast based on their width-to-length ratios.

Let us keep in mind the difference between ovals and ellipses. An oval is a curve resembling a squashed circle but it does not have a precise mathematical definition. The word *oval* is derived from the Latin word "ovus" for egg. Ovals sometimes have only a single axis of reflection symmetry instead of two. An ellipse is 1) a closed plane curve generated by a point moving in such a way that the sums of its distances from two fixed points is a constant, or 2) The shape resulting when a cone is cut by an oblique plane that does not intersect the base, also called a conic section.

An ellipse is defined by the equation:

$$\frac{x^2}{a^2} + \frac{y^2}{b^2} = 1$$

The following image shows bays near Bowmore, NC. Three of the bays have been overlaid with ellipses having the same width-to-length ratios as the bays. The fit of the ellipses is as perfect as could be expected for geological features. The outline of the ellipses fit precisely within the bay following exactly the borders of the bay. The remaining bays in this image can also be precisely fitted with elliptical equations.

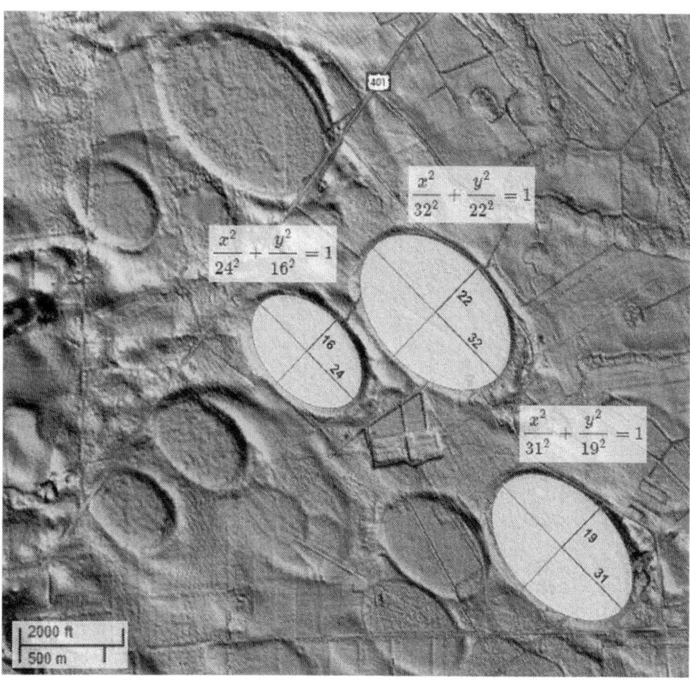

Figure 24. Bays near Bowmore, NC (Lat. 34.9196, Lon. -79.3206)

The ellipse fitting procedure is very simple. Select a potential Carolina Bay with well-defined boundaries and measure its width and length. Then, create an ellipse with a major axis proportional to the length and a minor axis proportional to the width. Rotate and scale the ellipse. For a Carolina Bay, the ellipse should fit exactly because Carolina Bays are conic sections.

The following image of Rainwater Basins in Nebraska shows that the well-formed bays can also be precisely fitted by ellipses, although there are many bays whose features have been highly degraded by geological processes. The orientation of the Nebraska bays is almost perpendicular to the bays in the East Coast.

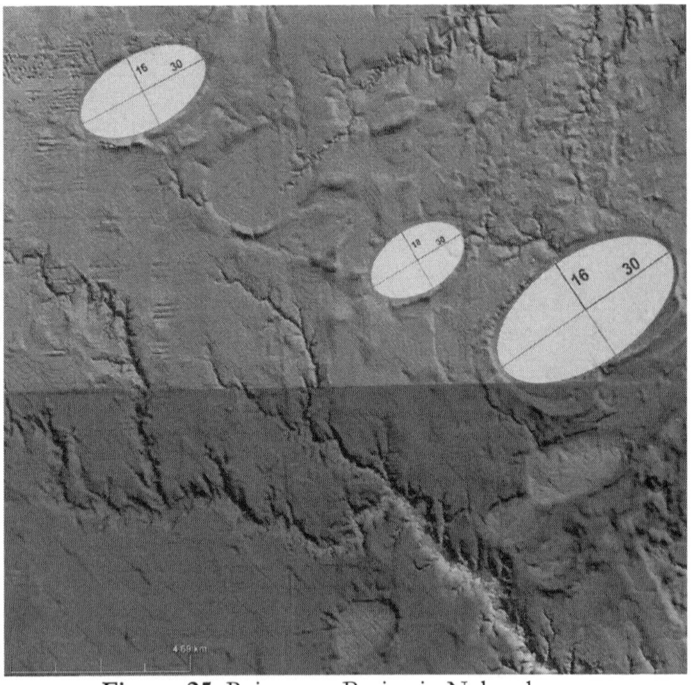

Figure 25. Rainwater Basins in Nebraska (Lat. 40.528, Lon. -98.084)

Having shown that the Carolina Bays and the Nebraska bays can be precisely fitted with ellipses, we can try to determine if there is a statistical relation between them. The following table was produced using *Google Earth* with the LiDAR data from Davias. Twenty-three bays with clearly defined borders were selected, 17 from the East Coast and 6 from Nebraska. The well-defined borders were necessary to be able to measure the major and minor axes accurately using Google Earth with a LiDAR overlay. The latitude and longitude correspond to the center of the bay.

Geometrical Characteristics of the Carolina Bays

Latitude	Longitude	Major axis meters (L)	Major axis meters (W)	W/L
34.372438	-79.955112	2875	1582	0.5503
34.355453	-79.897354	1142	627	0.5490
34.348281	-79.885369	829	509	0.6140
34.351854	-79.839793	1193	756	0.6337
34.341673	-79.805336	2773	1505	0.5427
34.354419	-79.792716	1234	812	0.6580
34.281692	-79.769796	2406	1362	0.5661
34.457433	-79.914511	1438	752	0.5229
34.437885	-79.876300	519	281	0.5414
34.387512	-79.831597	1216	780	0.6414
34.516371	-79.685701	2013	1208	0.6001
34.388400	-79.582470	1550	890	0.5742
34.384065	-79.524621	1478	867	0.5866
34.526274	-79.531788	1302	777	0.5968
34.527690	-79.485771	3125	1765	0.5648
34.623532	-79.579738	1848	1107	0.5990

34.644750	-79.637077	1422	871	0.6125
40.516923	-98.031921	5282	2854	0.5403
40.566439	-98.166159	3627	1940	0.5349
40.436968	-97.976410	1490	990	0.6644
40.518205	-99.135957	2496	1210	0.4848
40.425091	-98.985710	2150	1200	0.5581
40.763450	-97.927499	2660	1570	0.5902
Average				0.5794
Sample Standard Deviation				0.0452

The Carolina Bays and the Nebraska bays are both oriented so that their major axes intersect around the Great Lakes area, and in addition, they can both be fitted with ellipses with similar aspect ratios. The small standard deviation of the width-to-length ratios shows that there is a strong relationship between the Carolina Bays and the Nebraska bays, and this suggests that both were created by a common mechanism.

The consistent width-to-length ratios of the elliptical Carolina and Nebraska bays can be explained if the bays were originally conic sections formed from oblique conical cavities. The width-to-length ratios of the bays are related to the angle of the cone by the relationship: **$\sin(\theta) = W/L$**. The average 0.58 width-to-length ratio of the Carolina Bays corresponds to an angle of 35.5 degrees.

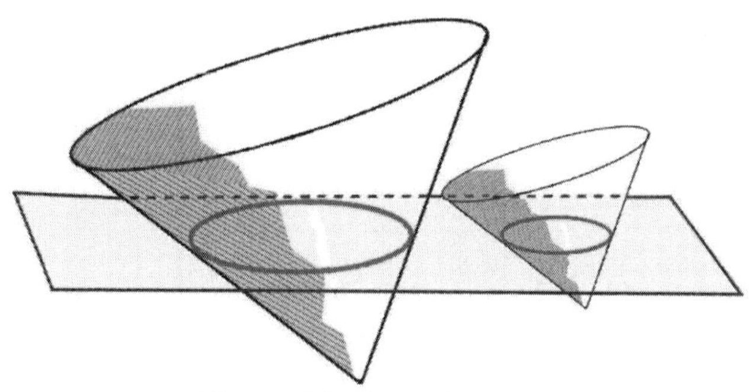

Figure 26. Conic Sections

Small and large ellipses have the same aspect ratios when the oblique cones have the same angle. We can postulate that the Carolina Bays and the Nebraska Rainwater Basins originated as oblique conical cavities slanted at an angle of approximately 35 degrees, and the cavities were remodeled later by geologic processes into elliptical bays. The large bays were made by large cavities and the smaller bays were made by smaller cavities. The small and the large bays have the same width-to-length ratios because they both were made from conical cavities at the same angle. The small variations of width-to-length ratios are due to slightly different angles of the conical cavities.

Not all bays are elliptical

Professor Douglas Johnson proposed a hypothesis of complex origin for the Carolina Bays that supposes that artesian springs, rising through moving groundwater and operating in part by solution, produced broad shallow basins occupied by lakes. Wave action created beach ridges at the margins of the lakes and dune ridges were formed by wind action (Johnson 1942). Most modern hypotheses are variations of these terrestrial mechanisms. This is what Prof. Johnson wrote regarding the shape of the bays:

"The perfection of curvature in oval outline observed in airplane photographs of the Carolina bays is so remarkable that one may hesitate to believe that the processes described in this volume are capable of producing such extraordinary results."

"It is easy to understand why anyone who has not made an intensive study of the bays should be strongly affected by this consideration. The photographs of bays selected for publication do have a perfection of outline which appears incredible as a product of those ordinary processes with which we are familiar. But the consideration would lose much of its force if the individual in question were to examine some thousands of photographs on file at various survey offices in an effort to find illustrations of comparable excellence."

Although the "perfection" of the shape was noted even when the first pictures of the Carolina Bays were published in the 1930s, no attempts were made then or subsequently to try to check the fit of the bays against geometric shapes such as ellipses. A visual or mathematical method could have been used to determine the divergence of the shape of the bays from the geometric figures. Instead, the shapes were completely disregarded and the eolian and lacustrine mechanisms became entrenched without an adequate explanation of how the precise elliptical shapes could have formed. Professor Johnson was correct in saying that it is hard to believe that wind and water terrestrial mechanisms could have created the elliptical bays because there is no geological analogy. Many places have artesian springs, wind, lakes and sandy soil, so why are such elliptical features not found anywhere else on the Earth? Even the most modern hypotheses for the formation of the Carolina Bays do not offer a computational model or a reasonable explanation of how these terrestrial mechanisms can create regular elliptical structures with specific aspect ratios in the two-thousand kilometer range from Nebraska to the east coast of the United States. This deficiency greatly weakens the credibility of the eolian and lacustrine hypotheses.

Prof. Johnson was convinced that, in spite of their great number, the elliptical bays were atypical features that formed fortuitously by ordinary geological processes because the aerial photographs that he examined also had many bays that differed in shape, and because earlier work had shown the tendency of lakes to acquire rounded or oval outlines when formed in deposits of uniform and limited resistance. The idea that a multitude of perfect ellipses can be formed by chaotic eolian and lacustrine processes is still very prevalent today. For this reason, geologists have not attached any particular significance to the precise elliptical shape of the Carolina Bays even though they occur only in the United States.

Modern LiDAR imagery shows bay boundaries that were not visible in twentieth century aerial photographs. The new images show a preponderance of elliptical bays that make it possible to propose that all the Carolina Bays were originally elliptical, but that terrestrial processes modified them after their formation into their current configuration. This agrees with Prouty's (1952) observations. It is also necessary to consider that some of the bays were never elliptical. There are two basic reasons why a bay could deviate from an elliptical shape:

- An elliptical bay was modified by geological processes after its formation.
- The terrain did not allow the formation of an inclined conical cavity.

Elliptical bays modified by terrestrial processes

The image of the Nebraska Rainwater Basins had some bays that were highly eroded by water or modified by windblown sediment. **Figure 27** of bays near Branchville, SC also shows some highly eroded bays. The whole landscape is covered with bays, but it is difficult to find one that is in pristine elliptical condition. Many of the bays can still be fitted with ellipses, although portions of the sandy rims have been breached or washed away. It is evident that these bays had a regular geometrical shape at some time in the past, but wind, water and surface movements have blurred and changed their original appearance over millennia. The original regular nature of these bays should not be dismissed just because their features have been worn away by wind and water erosion.

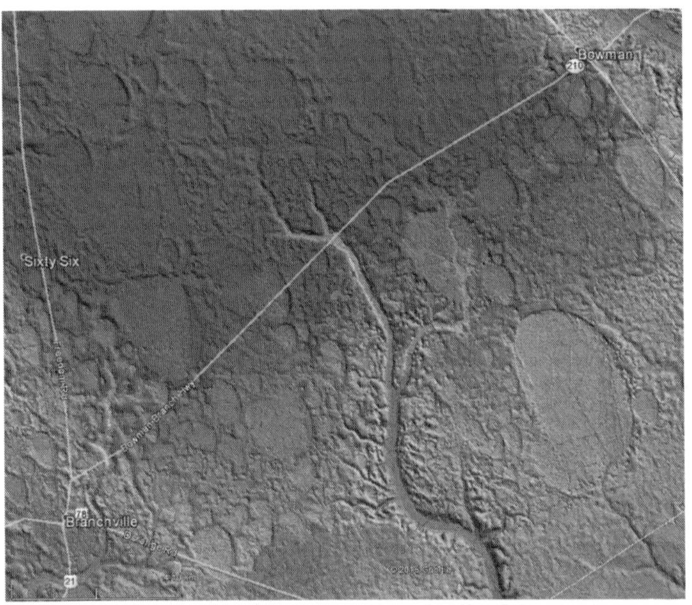

Figure 27. Bays near Branchville, SC
(Lat. 33.29947, Lon. -80.73608)

The Carolina Bays can undergo plastic deformations when the terrain on which they are located experiences ground movement. **Figure 28** shows two large bays in South Carolina that have been distorted in this manner. Hilson Bay has been stretched asymmetrically. Its northwest half is still rounded, but its southeast half has been pulled into a point toward a channel that drains into the Little Pee Dee River. Eleven kilometers to the south, Catfish Bay is in a trough that drains to the northwest into the Great Pee Dee River and to the southeast into the Little Pee Dee River. Consequently, Catfish Bay has been stretched in both directions giving it an atypically low width-to-length ratio. Even though these bays are now distorted, they were probably elliptical when they originally formed. Deviations of the bays from their prototypical elliptical shape may prove useful in determining the historical rate of horizontal ground movement.

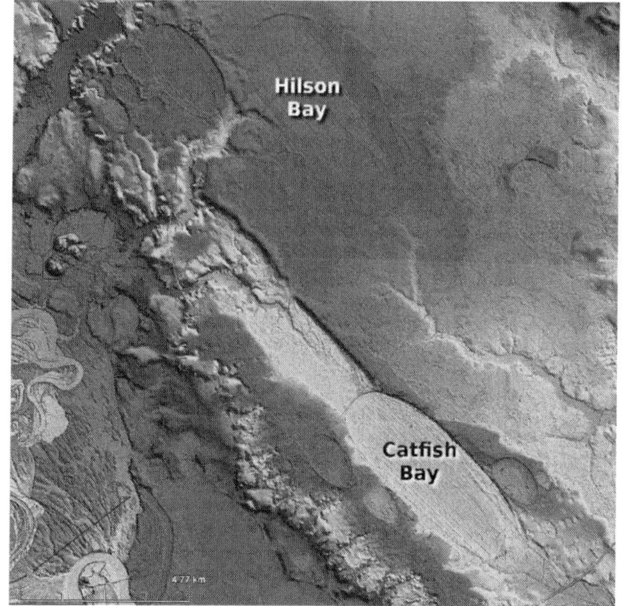

Figure 28. Hilson Bay (Lat. 34.47519, Lon. -79.58657) and Catfish Bay (Lat. 34.38312, Lon. -79.54918)

Terrain unsuitable for the formation of conical cavities

The variability of the terrain on which the bays were formed influenced their shape. Ground that could not be liquefied or that had only a shallow layer of unconsolidated material could not have produced conical cavities, and the resulting bays, if any, could not have been elliptical. The thickness of the layer of unconsolidated material required to produce an elliptical bay can be estimated by the formula $\tan(\theta) \times L/2$, where L is the length of the major axis and θ is the angle of inclination. Thus, the formation of a conical cavity inclined at 35 degrees corresponding to a bay with a major axis of 400 meters would require a layer of unconsolidated material with a depth of approximately 140 meters.

Impacts on viscous media can consistently make conical cavities. Upon hitting a viscous surface, a projectile parts the medium and slows down, finally coming to rest at the bottom of a conical cavity. But what happens when the target surface does not have a deep enough layer of unconsolidated material? In this case, the initial stage of the impact parts the medium as before, but the projectile disintegrates when it is stopped by the solid barrier, and its energy is released explosively. The resulting cavity is hemispherical rather than conical, as will be discussed in a later chapter. Davias (2015) has identified six types of bay archetype shapes that may be determined by the geological characteristics of the terrain.

The calculations made on the basis of the elliptical shapes of the bays are not invalidated when we understand the reasons why some bays differ from their presumed prototypical shape. When the Carolina Bays become the subject of rigorous morphological inquiry, it will be important to take into consideration the constitution of their substrata to account for deviations from their expected shape.

Figure 29 shows Carolina Bays near Barclay, Maryland on the Delmarva peninsula. The surface is pockmarked by highly eroded bays and none of them appear to be elliptical. This would be expected for terrain with insufficient depth of unconsolidated material.

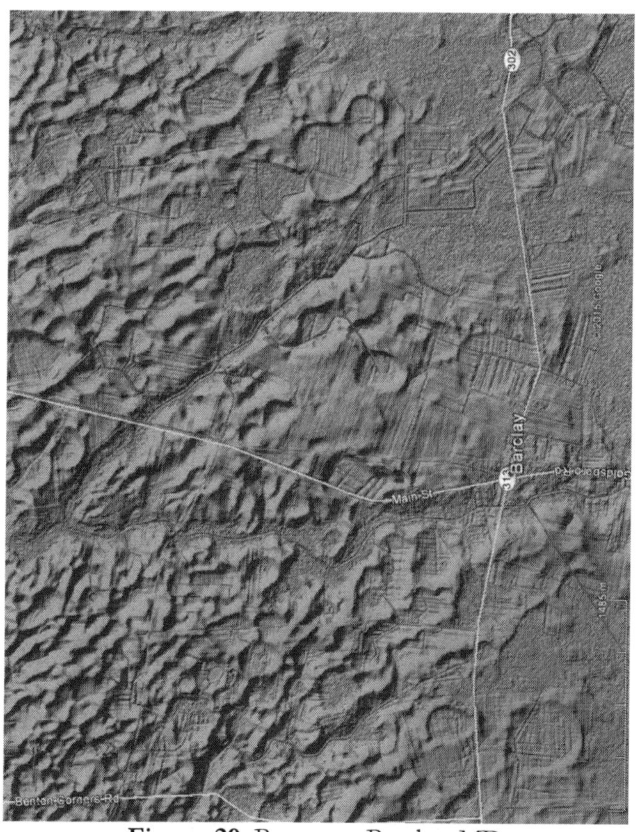

Figure 29. Bays near Barclay, MD
(Lat. 39.15933, Lon. -75.85541)

* * *

ALIGNED LAKES

The previous chapter demonstrated that the well-formed Carolina Bays and the Nebraska Rainwater Basins could be fitted precisely by ellipses with corresponding width-to-length ratios. This is a fundamental property resulting from the mechanism by which the bays were formed.

The Carolina Bays have been characterized as being eolian blowouts (Zanner and Kuzila 2001) and as thaw lakes formed when permafrost melts (Melosh 2011). Thermokarst lakes in Alaska and Russia form when underground ice melts in a region underlain by permafrost and the ground collapses like a sinkhole. The cavity fills with water forming a lake, but unlike the Carolina Bays, the lakes are not all perfect ellipses, they have no overlaps, they have no preferential rim thickening, and they do not have raised rims. The shape and alignment of the thaw lakes is determined by the contours of the land, and the alignment is generally in the direction in which water drains toward lower terrain. We can show that the properties of such thermokarst lakes are completely different from the Carolina Bays by using the same ellipse-fitting procedures and statistical methods that were used for the bays.

Google Earth makes it possible to zoom in and measure the width and length of the lakes. **Figure 30** of aligned lakes in Alaska illustrates that an ellipse with the same width and length as one of the lakes is not a good fit. Some of the margins of the lake fall outside of the ellipse, and some of the margins are inside of the ellipse. A few thermokarst lakes in Alaska may be elliptical, but that is an exception, rather than the rule. In general, thaw lakes are not elliptical and should not be used as models for the Carolina Bays.

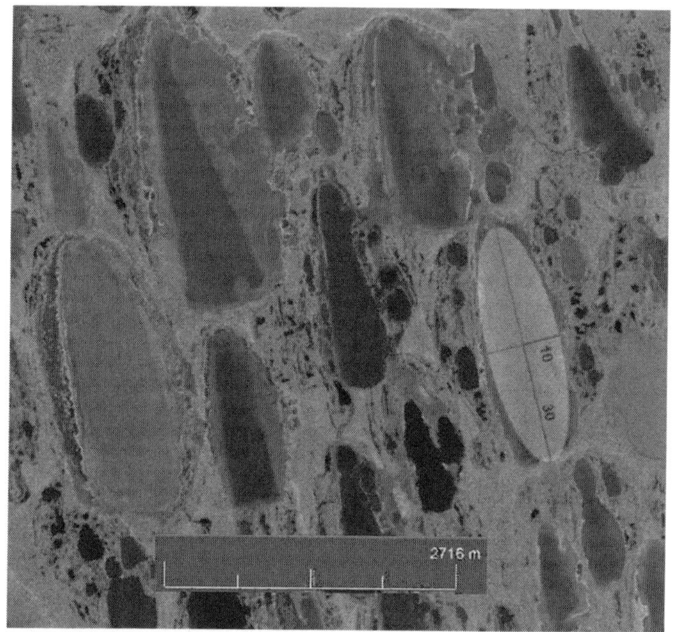

Figure 30. Aligned lakes in Alaska thermokarst are not elliptical (Lat. 70.2894, Lon. -158.9830)

The following LiDAR image of Alaska's north coast by the Beaufort Sea (**Figure 31**) shows a wide variety of aligned lakes with no consistency in their shapes or aspect ratios. The LiDAR images of the Carolina Bays show smoothly curved edges, but the edges of the Alaskan aligned lakes are rough and lack precise curvatures.

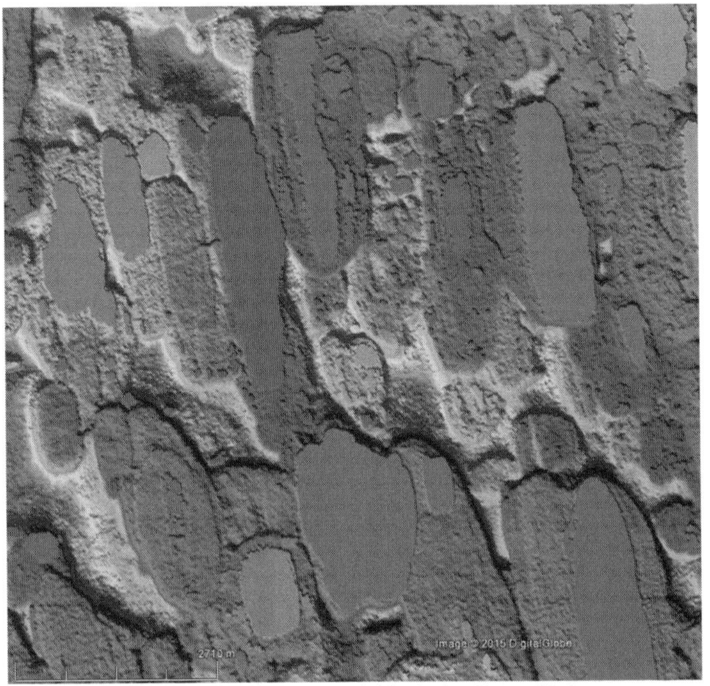

Figure 31. LiDAR image of Alaska's aligned lakes (Lat. 70.81338, Lon. -157.49498)

Visual comparison indicates that the aligned lakes and the Carolina Bays must have had different mechanisms of origin. The distinction between Carolina Bays and thermokarst lakes can be further demonstrated statistically by tabulating the width-to-length ratios. **Figure 32** shows a group of 15 lakes that have been assigned identifying tags.

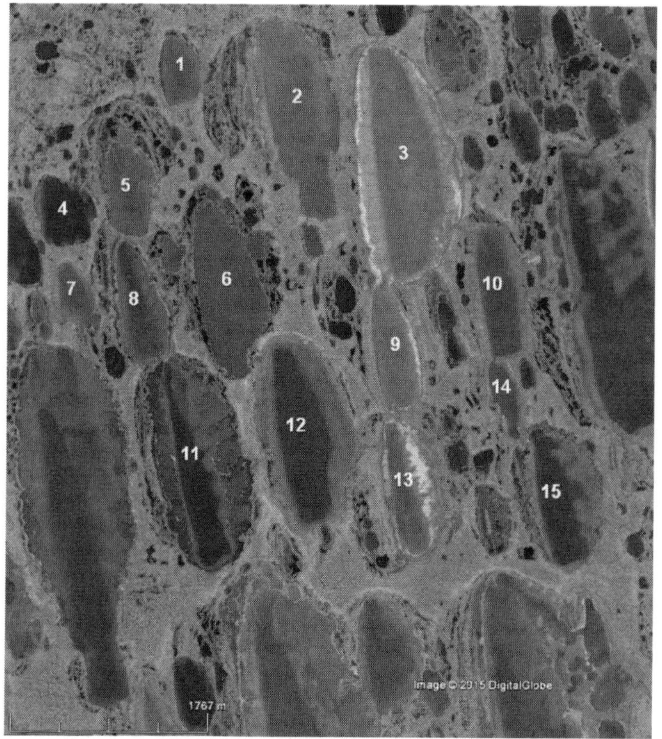

Figure 32. Alaska lakes numbered for analysis (Lat. 70.3358, Lon. -159.0568)

The following table has the width-to-length ratios of these 15 Alaska thermokarst lakes in the vicinity of the coordinates Lat. 70.3358, Lon. -159.0568.

The statistical results show that the thermokarst lakes in Alaska are significantly different from the Carolina Bays. The average width-to-length ratio for the Alaska lakes is 0.45, which is quite different from the 0.58 for the bays. In addition, the standard deviation is 2.6 times greater than for the bays. The standard deviation for the lakes (0.119) indicates great variability in the shapes of the lakes, whereas the shape of the Carolina Bays is more consistent.

Width-to-length ratios of thermokarst lakes			
Tag	**Width (meters)**	**Length (meters)**	**W/L**
1	326	555	0.5874
2	590	1762	0.3348
3	864	2061	0.4192
4	479	610	0.7852
5	425	847	0.5018
6	579	1552	0.3731
7	234	540	0.4333
8	387	992	0.3901
9	372	1048	0.3550
10	328	1039	0.3157
11	820	1836	0.4466
12	823	1677	0.4908
13	532	1137	0.4679
14	220	581	0.3787
15	664	1255	0.5291
		Average	0.4539
		Standard Deviation	0.1190

Other structures that have been suggested as analogous to the Carolina Bays are the aligned salt lakes in Australia. These salt lakes form when marshy ground dries up, but this mechanism does not create the raised rims or overlapping ellipses that are characteristic of the Carolina Bays. The following is an image of the Australian salt lakes. A similar ellipse-fitting and width-to-length statistical analysis can be used to show that these salt lakes are quite different from the Carolina Bays.

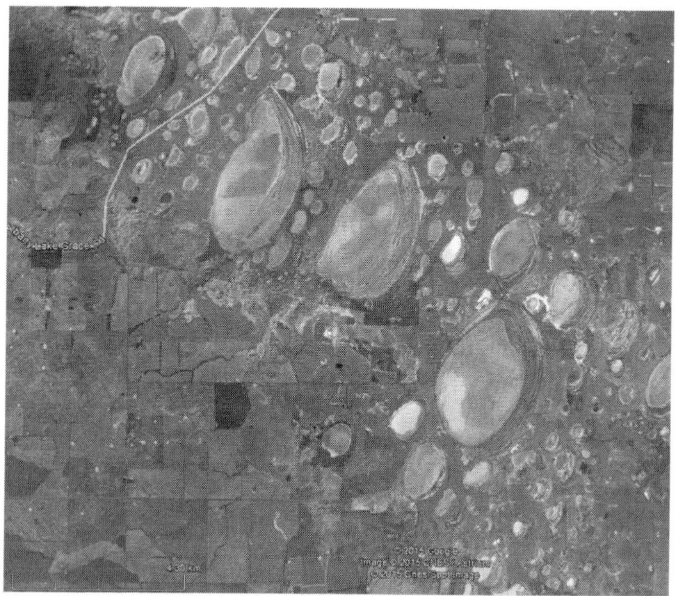

Figure 33. Salt lakes in Australia
(Lat. -33.6051, Lon. 118.5143)

The statistical analysis of the physical proportions of several geological features makes it possible to differentiate thermokarst and salt lakes from the Carolina Bays while at the same time emphasizing the similarities that exist between the Carolina Bays and the Nebraska Rainwater Basins. Geometrical and statistical analysis is usually not used to characterize geological features because of their great variability, but it can be very useful for discovering patterns that might not be obvious from visual inspection, such as the precise fit of elliptical equations to the Carolina Bays. It is significant that the aligned lakes are not oriented toward an epicenter like the Carolina Bays and the Nebraska Rainwater Basins.

* * *

THE YOUNGER DRYAS IMPACT HYPOTHESIS

The impact origin of the Carolina Bays has been a subject of bitter controversy for many years. The main point of disagreement is the type of evidence that can be reliably used to claim that an impact occurred sometime in the past. The burden of proof for an impact is very rigorous. In the case of the impact that killed the dinosaurs 65 million years ago, geologists found traces of iridium at the Cretaceous–Paleogene (K-T) boundary where the dinosaurs disappeared (Alvarez 1980). Iridium is a metal that is more common in meteorites than on the Earth's crust. More than ten years passed before shock-metamorphic materials were found in drill samples from an oil-exploration project in an undersea crater near the Yucatan peninsula in Mexico (Hildebrand 1991). The K-T extinction impact hypothesis was generally accepted only after the petrographic evidence had confirmed the occurrence of a powerful impact.

In 2007, Richard B. Firestone and several co-authors proposed that an extraterrestrial impact 12,900 years ago had contributed to the extinction of the North American megafauna and to the Younger Dryas cooling period (Firestone 2007). This proposal was later nicknamed "The Younger Dryas Impact Hypothesis." The time of the proposed impact was during the ice age when the northern United States and Canada were covered by the Laurentide Ice Sheet, which was hundreds of meters thick, and large animals such as saber-tooth tigers, camels and mammoths roamed the land. Humans, such as the Clovis people, had also colonized America. The greatest concentrations of Paleoindian fluted arrowheads and spear points are found east of the Mississippi River, although the earliest fluted point sites are clustered in the Southern Plains in Texas, Oklahoma and New Mexico (Anderson 2009).

Younger Dryas cooling period

Approximately 14,700 years ago, a prolonged warming period opened an ice-free corridor across Canada. The warming period was interrupted abruptly approximately 12,900 years ago with the onset of the Younger Dryas cooling period (**Figure 34**). This cold interval that lasted approximately 1300 years was first recognized in European pollen records that showed the reappearance of a cold-tolerant Arctic flowering plant (*Dryas octopetala*) from which the name of this period was derived. With the onset of this cold period, temperatures in the Northern Hemisphere suddenly returned to near-glacial conditions.

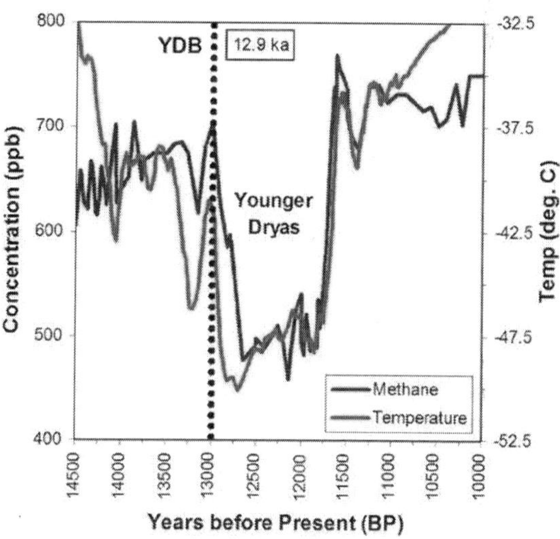

Figure 34. Ice cores in Greenland provide temperature records. (Modified from Firestone)

Because no crater site had been found, Firestone's 2007 paper proposed that one or more low-density extraterrestrial objects exploded over northern North

America, partially destabilizing the Laurentide Ice Sheet and triggering the Younger Dryas cooling. Ten Clovis-era sites were examined for evidence, and the evidence consisted of a discrete layer with magnetic microspherules, magnetic grains with iridium, charcoal, soot, carbon spherules, glass-like carbon containing nanodiamonds and other indicators of large biomass burning. The layer was also found throughout at least 15 Carolina Bays.

Since there was no impact site and no petrographic evidence, Firestone's paper was not well received. In Firestone's own words, the results "unleashed an avalanche of controversy." The 2007 paper mentioned the Carolina Bays only as the location from where samples were taken, but without making a determination whether any bays were or were not formed by the Younger Dryas event.

In 2009, Firestone wrote another paper reiterating the case for a Younger Dryas extraterrestrial impact to account for the extinction of the mammoth, megafauna and Clovis people 12,900 years ago (Firestone 2009). The paper emphasized research by others that supported his earlier claims about magnetic microspherules, nanodiamonds, and the chemical composition of the evidence. There was also more discussion about the extraterrestrial impact and the formation of the Carolina Bays.

Firestone's paper included a map of the United States and Canada with some images showing the radial orientation of the Carolina Bays and lines linking the Great Lakes and Hudson Bay to the location of the bays. Blue lines were drawn radiating from the Great Lakes toward Virginia, North Carolina, South Carolina, Georgia, Nebraska, Kansas, New Mexico and Texas. Red lines were drawn from Hudson Bay to the East Coast states.

According to Firestone, the high concentrations of water in the magnetic grains was consistent with an airburst occurring over the Laurentide Ice Sheet that would have sent a high-temperature shock wave that created the Carolina Bays as it passed. Firestone also suggested that an impact on the ice might have left a crater that could easily be hiding within the Great Lakes where the action of water rushing out of the failing glacier would have erased many of its features.

In answering criticism about the Younger Dryas Impact Hypothesis, Firestone noted that the variations in the dates of the Carolina Bays could reflect inadvertent sampling of underlying, older sediment that may have shifted over time. He emphasized that the strikingly regular orientation of the Bays with their major axes pointing toward the Great Lakes or Hudson Bay was inconsistent with their formation during major Atlantic storms under variable wind conditions, but was consistent with their formation by a shockwave coming from the Great Lakes, although it was also likely that the Bay contents had shifted and mixed with newer sediments over time by the action of wind and water. A report about the occurrence of nanodiamonds in the Younger Dryas boundary sediment layer appeared the same year (Kennett 2009), but it did not result in increased general support for the Younger Dryas Impact Hypothesis.

The Requiem Paper

In 2011, Nicholas Pinter from Southern Illinois University and six other eminent academic scientists wrote a rebuttal paper entitled "The Younger Dryas impact hypothesis: A requiem". The title made it clear to the scientific community that this topic was dead and that scientists should not bother to consider additional evidence in support of this discredited impact hypothesis. The requiem paper refuted all of Firestone's evidence point-by-point and chastised him and his colleagues for not presenting "recognized and expected impact markers" and for proposing "impact processes that were novel, self-contradictory, rapidly changing, and sometimes defying the laws of physics."

The paper presented scientific arguments against the Younger Dryas Impact Hypothesis and the methodology used by its proponents. Pinter's paper listed overwhelming evidence that none of the impact markers presented by Firestone and his associates could be used reliably to prove that there had been an airburst or impact of an extraterrestrial object. The micrometeorite particles and extraterrestrial helium could not be substantiated. The report of increased iridium could not be reproduced. The Carolina Bays did not have meteoritic material and their dates were too diverse to have been caused by a single event. The nanodiamonds could have been deposited by various sedimentary processes and were not necessarily attributable to an extraterrestrial event, and finally, the spherules could be fungal structures or termite dung.

The requiem paper warned the scientific community to adhere to a rigorous approach using unambiguous criteria for extraterrestrial impacts and avoid being too eager to jump aboard the "impact bandwagon" when confronted with unusual geological evidence.

The impact origin of the Carolina Bays was placed into the category of evidence that "has been largely rejected by the scientific community and is no longer in widespread discussion." The requiem paper pointed out that Firestone had suggested two completely different mechanisms for the formation of the Carolina Bays. A book by Firestone et al. (2006) had implied that the Carolina Bays formed from impacts of large-scale secondary ejecta from the primary impact site, but Firestone (2009) suggested that the bays had formed from a high-temperature shock wave. Pinter and his colleagues emphasized that the Carolina Bays had not formed instantaneously, but rather over significant time, and provided references to independent dating studies showing "multiple periods of bay-rim accretion with intervening intervals of erosion."

Aftermath of the requiem paper

As expected, funding for the Younger Dryas Impact Hypothesis dried up and the mere mention of the Carolina Bays put scientists on their guard. Many journals stopped accepting manuscripts on these topics, but some papers about the impact spherules managed to get in print, such as LeCompte (2012), Israde-Alcántara (2012) and Wittke (2013).

Research on the Greenland ice cores revealed a surprise. In 2013, Michail I. Petaev and his colleagues from Harvard University published a paper announcing that they had found a platinum anomaly in the Greenland ice sheet that pointed to a cataclysm at the onset of Younger Dryas. The authors concluded that the combination of chemical compounds found in the Greenland glacier ice core hinted at an extraterrestrial source of platinum, possibly from an iron meteorite of low iridium content, and that such a meteorite was unlikely to result in an airburst or trigger wildfires over large areas of America.

The extraterrestrial impact proposed by Petaev makes it possible to consider an extinction event again. However, the diverse dates obtained for the Carolina Bays still pose a large obstacle for displacing the well-established wind and water formation hypothesis in favor of an impact origin that has been largely rejected by the scientific community. Very convincing and irrefutable proof will be needed to show that the Carolina Bays were created by impacts and that all those very accurate dates do not correspond to the time of formation of the bays.

* * *

THE GLACIER ICE IMPACT HYPOTHESIS

It may seem foolhardy to propose an impact hypothesis for the Carolina Bays considering that the mainstream scientific community has largely rejected this idea because of the wide diversity of dates for the bays and a large body of geological studies documenting multiple periods of bay-building processes. Similarly, any claim that the megafaunal extinction and the Younger Dryas cooling period were caused by an extraterrestrial impact for which no crater has been found will also be met with great skepticism, especially after the rejection of the Younger Dryas Impact Hypothesis (YDIH).

The YDIH was developed in a top-down approach: The premise that there was a cosmic explosion was followed by efforts to find physical evidence to substantiate the hypothesis. Unfortunately, the proponents of the hypothesis could not produce the type of evidence expected by impact scientists.

The only evidence for an impact at the Younger Dryas boundary is the platinum anomaly reported by Petaev (2013) and information on impact spherules such as reported by Wittke (2013), but this information is not enough to establish an extinction event.

A New Hypothesis

The fact that the Carolina Bays and the Nebraska Rainwater Basins are true ellipses and that extensions of the bays' major axes intersect in the Great Lakes region demand the formulation of a new hypothesis about the origin of the Carolina Bays, particularly because the wind-and-water hypotheses fail to explain these two properties of the bays.

The relationship between ellipses and cones is a fundamental mathematical principle. The new hypothesis starts from the observation that the Carolina Bays are conic sections with width-to-length ratios of about 0.58, and therefore, it is reasonable to postulate that the bays originated from oblique conical cavities slanted at an angle of approximately 35 degrees and that the cavities were remodeled into shallow elliptical bays by geological processes. As will be shown later, conical cavities can be consistently produced by impacts on viscous targets.

Another essential component of the new hypothesis is the observation that the point at which the axes of the bays intersect in the Great Lakes region was covered by the Laurentide ice sheet with a thickness of approximately 1500 to 2000 meters (Dyke, 2002) during the Pleistocene Epoch. This suggests that ice chunks from this point of origin could have been the projectiles that formed the conical cavities for the bays. A meteorite impact on the glacier is the only natural process that could have launched chunks of ice at an angle of approximately 35 degrees in ballistic trajectories with a range of 1500 kilometers. Using a bottom-up approach based on the physical evidence, we have deduced that a meteorite impact was necessary for the formation of the Carolina Bays. The new hypothesis will be called the Glacier Ice Impact Hypothesis (GIIH). Its goal is to elucidate the physical mechanisms of the extraterrestrial event, the glacier ice impacts, and the geologic process that converted the conical cavities into Carolina Bays.

The Glacier Ice Impact Hypothesis was first described before the geometrical survey had demonstrated the perfect elliptical shape and statistical relationship of the Carolina Bays and the Nebraska Rainwater Basins (Zamora 2013, 2014). The Glacier Ice Impact Hypothesis depends on four different physical processes that must have occurred in a specific sequence in order for the

Carolina Bays to have been created by secondary impacts from material ejected by a primary extraterrestrial impact. The hypothesis provides two different ways of proving that the Carolina Bays were created by impacts even if the location of the primary impact is never found.

The four main processes of the Glacier Ice Impact Hypothesis in their order of occurrence are:

1. An asteroid or comet impact on the Laurentide ice sheet ejected glacier ice boulders of various sizes at ballistic speeds.
2. The unconsolidated ground in the target areas was liquefied by the seismic shock waves from the primary and secondary impacts.
3. Oblique impacts of glacier ice boulders on the liquefied ground created slanted conical craters.
4. Viscous relaxation reduced the depth of the conical cavities and remodeled them into elliptical bays.

The following paragraphs provide a brief outline of information that supports each of these four processes of the hypothesis. The information can be calculated from the location and size of the Carolina Bays or from geological data and the experimental results of a physical model.

1) Meteorite Impact on the Laurentide ice sheet

The size of the Carolina Bays and the distance from their point of origin can be used as follows:

- The launch speed and the trajectory of an ice projectile can be calculated with ballistic equations using the distance of each bay from the point of origin.
- The size of a bay can be used to estimate the energy of the impact and the size of the projectile by using yield-scaling laws.

- From the number of bays and the average size of the projectiles, it is possible to estimate the minimum volume of ice ejected by the extraterrestrial impact and its minimum energy.

2) Soil liquefaction by seismic shock waves

The liquefaction of saturated soil by earthquakes is a well-known phenomenon that can topple buildings and cause cars to sink.

- The estimates of projectile impact energy can be used to determine their effectiveness for causing soil liquefaction.
- The geology of the terrain, such as the depth to the water table in the eastern seaboard and Nebraska, can provide clues about the formation of viscous ground suitable for the creation of conical cavities.

3) Impacts on viscous targets create conical craters

Conical cavities created by impacts are usually transient. The physical attributes of the target surface determines how fast the conical cavities are remodeled by elastic or gravitational forces.

- Impacts on viscous surfaces create conical cavities with raised rims.
- Some impacts on viscous surfaces create overturned flaps.
- Impacts on viscous surfaces part the target without destroying its stratigraphy
- Oblique impacts create slanted conical craters.
- Oblique impacts may have a thick rim of material pushed in the direction of projectile travel.

4) Viscous relaxation reduces cavity depth

Gravity reduces craters and peaks to the ultimate limit of stability: a level plain. The mechanics of plastic deformation and viscous flow can be used to explain how oblique conical craters turned into elliptical bays.

- Viscous relaxation decreases the depth of a cavity from the bottom up.
- Viscous relaxation restores the original stratigraphy.
- Only the surface of a conical crater is exposed to light during the impact and remodeling thereby placing constraints on the methodology for dating the subsoil.

How to test the Glacier Ice Impact Hypothesis

One of the objections to the Younger Dryas Impact Hypothesis was that it did not present recognized impact markers for an extraterrestrial event. The Glacier Ice Impact Hypothesis only needs to prove that the Carolina Bays originated from conical cavities caused by ballistic impacts of glacier ice. Once that is established, it will be logical to conclude that there must have been an extraterrestrial impact on the glacier ice sheet, even though it may be difficult or impossible to determine exactly where.

Ballistic impacts on viscous ground do not generate the kind of pressures necessary to create shock metamorphism at the impact site. In addition, terrestrial glacier ice does not have any siderophile elements that would leave a characteristic signature. However, two tests can be performed on the Carolina Bays to prove that they are the remodeled remnants of conical impact cavities made by glacier ice. 1) All impact cavities have raised rims and may have overturned flaps with inverted stratigraphy. Finding inverted stratigraphy in the raised rims of the Carolina

Bays would offer proof of their impact origin. 2) Glaciers typically contain embedded stones from the terrain on which they form. Finding stones from the Great Lakes region at the bottom of the conical craters where the glacier ice boulders stopped would also confirm the impact origin of the Carolina Bays. Once it is established that the Carolina Bays are impact structures, land features in the form of geometric conic sections may gain acceptance from geologists and astronomers as a signature of impacts on viscous surfaces.

The next chapters will examine the four main processes of the Glacier Ice Impact Hypothesis in more detail.

* * *

EXTRATERRESTRIAL IMPACT AND BALLISTIC FLIGHT

Although no crater from an extraterrestrial impact has been found associated with the megafauna extinction or the Younger Dryas cooling event, we can attempt to establish whether there was an extraterrestrial impact through mathematical forensic techniques with data derived from the Carolina Bays.

We start by using the distance of the Carolina Bays from the intersection point of the extensions of their major axes, which is presumably the location of the extraterrestrial impact and the point of origin of the glacier ice chunks that made the bays. The distance enables us to calculate the ballistic trajectories of the ejected ice. One of the compelling aspects of the Glacier Ice Impact Hypothesis is that it uses well-established principles of physics, such as Newton's laws of motion and the scaling laws relating yield energy to crater size. The mathematical framework of geometry and physics makes it possible to derive the conditions under which the Carolina Bays could have been created.

Ballistic Trajectories

The launch speed and the trajectory of an ice projectile can be calculated with ballistic equations using the distance of each bay from the point of origin (**Figure 35**). We can calculate that an ice boulder ejected at an angle θ of 35 degrees from Michigan to the South Carolina seashore would require a launch speed **v** of approximately 3.6 km/sec to cover the distance **D** of 1,220 kilometers using the following ballistic equation, where **g** is the acceleration of gravity.

$$D = (v^2/g)\sin(2\theta)$$

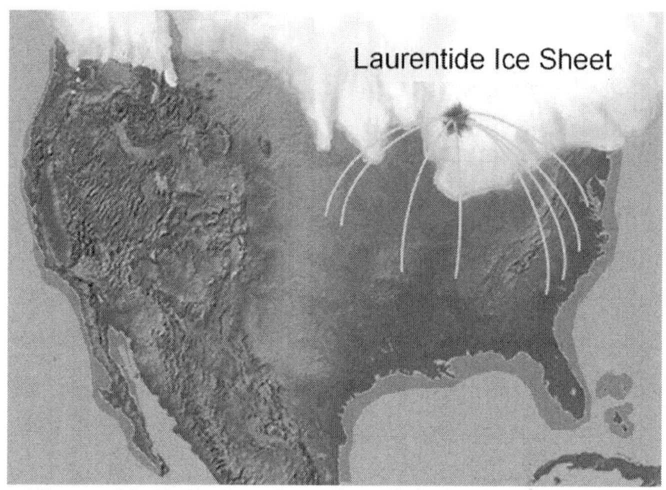

Figure 35. Ballistic trajectories of glacier ice ejecta after an extraterrestrial impact on the Laurentide ice sheet

At this speed, which is almost eleven times the speed of sound, the ice boulder would reach its target about 7 minutes after launch. The trajectory would take the boulder 213 kilometers above the surface of the Earth, which is more than a hundred kilometers above the atmosphere. The path of an object launched from Earth that reaches 100 kilometers above sea level, and then falls back to Earth, is considered a sub-orbital spaceflight. The time of flight **T** and the maximum height **H** of the trajectories are given by the equations:

$$T = (2v/g)\sin(\theta)$$

$$H = v^2\sin^2(\theta)/2g$$

Depending on the launch speeds and launch angles, some ice chunks would have reached heights of 150 to 390 kilometers above the surface of the Earth. The following table shows that the ballistic speeds would be in the range of 3 to 4 km/sec, depending on the distances to the target. Notice that all the flight paths go above the Earth's atmosphere; this will be important in the discussion about liquid ejecta.

$D = (v^2/g)\sin(2\theta)$

$T = (2v/g)\sin(\theta)$

$H = v^2\sin^2(\theta)/2g$

Distance (km)	Launch angle	Launch speed (km/sec)	time of flight (min)	max. height (km)
1,000	45	3.1	7.5	229
1,000	35	3.2	6.3	150
1,470	45	3.8	9.1	368
1,470	35	3.9	7.6	255

Figure 36. Characteristics of the ballistic trajectories

Projectile Sizes

Knowing that the glacier ice boulders had speeds in the range of 3 to 4 km/sec, we can now estimate the size of the boulders using yield-scaling laws that relate the size of a crater to the energy of the projectile. The mathematical foundation for correlating crater size to projectile size is provided in Melosh (1989). The University of Arizona has an online program for Computing Projectile Size from Crater Diameter (Melosh and Beyer, 1999). The size of a crater depends on the projectile's size, speed, and the angle at which it strikes. Factors such as the projectile's composition and the material and the structure of the target surface are also important.

The calculation in **Figure 37** estimates the size of an ice boulder that could make a Carolina Bay with a diameter of one kilometer on unconsolidated ground.

Computing Projectile Size from Crater Diameter

http://www.lpl.arizona.edu/tekton/crater_p.html

H. Jay Melosh and Ross A. Beyer

Results for computing projectile size from transient crater diameter

Your Inputs:

Final Crater Diameter **1 kilometer**
Projectile Density **917 kg/m³ (Ice)**
Impact Velocity **3 km/sec**
Impact Angle **45 degrees**
Target Density **1500 kg/m³**
Acceleration of Gravity **9.8 m/sec²**
Target Type **loose sand**

Results:

The three scaling laws yield the following projectile diameters:
(note that diameters assume a spherical projectile)
Yield Scaling 1.79×10^2 meters
Pi Scaling (Preferred method!) **180 meters**
Gault Scaling 2.61×10^2 meters
Crater Formation Time 6.56 seconds
Using the Pi Scaling method this impactor would have struck
the target with an energy of **1.27×10^{16} Joules (3.03 MegaTons)**.

Crater Program, Copyright© 2002 Ross A. Beyer & H. Jay Melosh
These results come with ABSOLUTELY NO WARRANTY.

Figure 37. Yield Scaling Calculation

According to the calculation, a crater with a diameter of 1 kilometer could be made by a spherical ice boulder with a diameter of 180 meters traveling at a speed of 3 km/sec when impacting at an angle of 45 degrees. The energy of the impact would be approximately 1.27×10^{16} Joules or 3.03 megatons of TNT explosive force. Each such impact would be the equivalent of a magnitude 7.54 earthquake. These figures should be considered approximations, since the theoretical work that has been done on scaling oblique impacts is based on noncohesive quartz sand (Gault and Wedekind 1978).

Using the formula for the volume of a sphere, an ice boulder with a diameter of 180 meters would have a volume of 3 million cubic meters and would weigh 2.8 million metric tons. The ice boulder would be about the size of Yankee Stadium in New York City.

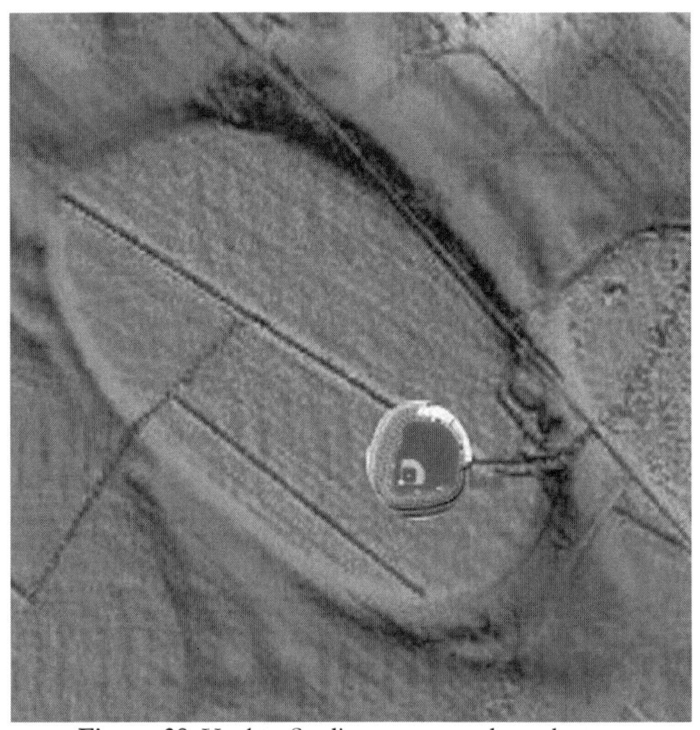

Figure 38. Yankee Stadium compared to a bay in North Carolina (Lat. 34.8418, Lon. -79.2166)

Figure 38 shows Yankee Stadium in proportion to a bay in North Carolina (Lat. 34.8418, Lon. -79.2166) with a major axis of approximately 942 meters. The size of the playing field and all the bleachers of Yankee stadium measure approximately 180 meters in diameter; this is about the same size as the glacier ice boulder that made the bay.

The Extraterrestrial Impact

From the number of bays and the average size of the projectiles, it is possible to estimate the volume of ice ejected by the extraterrestrial impact and its energy. If there are 500,000 Carolina Bays and each one was formed by an energy of 1.27×10^{16} Joules, then the total energy of the impacts was approximately 6.35×10^{21} Joules. This total energy provides a rough lower limit of the kinetic energy transferred to the ejecta by the extraterrestrial impact. Additional energy would have been converted to heat, seismic energy and fracturing of the target and projectile. The kinetic energy of 6.35×10^{21} Joules would correspond to a stony asteroid with a diameter of 3 kilometers or to a faster icy comet with a diameter of 2 kilometers.

Using the same number of bays, we can calculate that approximately 1.5×10^{12} cubic meters of ice were ejected by the impact on the Laurentide ice sheet; that is enough ice to cover half of the United States to a depth of half a meter. This amount of ice is also a minimum since much of the ice may have fallen on ground too firm for the formation of bays. If the ice sheet had a thickness of 1 kilometer at the point of the extraterrestrial impact, the circular area containing this volume of ice would have had a diameter of 44 kilometers.

Thus far, we have obtained an estimate of the energy of the extraterrestrial impact assuming that the Carolina Bays and the Nebraska Rainwater Basins were created by secondary impacts of glacier ice. Does this make sense from what we know about geology and extraterrestrial impacts?

The first premise of the Glacier Ice Impact Hypothesis is that an asteroid or comet impact on the Laurentide ice sheet ejected glacier ice boulders. Although no impact site has been found, the trace of platinum that Petaev (2013) found at the Younger Dryas boundary in the Greenland ice supports the idea that an extraterrestrial object could have had the chemical composition to survive its passage through the atmosphere and actually hit the ice sheet instead of just disintegrating in an airburst (**Figure 39**).

Figure 39. An extraterrestrial impact on the Laurentide Ice Sheet would have ejected ice, water and steam.

What happens when an extraterrestrial object hits a glacier? We know that an impact on a rocky surface generates a lot of heat, melts, and ejects pieces of rock. Something similar happens with ice. Ice is brittle and a bad conductor of heat (Schulson, 1999). Peter H. Schultz from Brown University conducted experiments with NASA's Ames Vertical Gun showing that ice shatters upon impact (Schultz 2009). Pieces of ice are ejected at high velocity radiating from the impact site. This radial distribution of ejecta is characteristic of impacts.

Since ice is three times less dense than rock, a meteorite impact on ice would encounter less resistance during the contact and compression stage than it would if it impacted rock. The shock wave of the impact would cause the ice to break up as the meteorite penetrated the glacier ice. The heat from the impact would produce great quantities of water, and a plume of steam at great pressure would help to propel the ice pieces during the excavation stage. The ejecta produced by the impact would be mainly ice, water and steam. If the meteorite penetrated all the way through the one- or two-kilometer thickness of the Laurentide ice sheet, then additional rocky material might be ejected. During the modification stage, a large quantity of melt water would flood the impact site, creating lakes in the depressions and perhaps washing away evidence of the extraterrestrial impact. Any siderophile elements associated with the extraterrestrial impact might end up in aquatic sediments downstream from the impact site. The ice boulders ejected during the impact would travel above the atmosphere and blanket an area with a radius of approximately 1500 kilometers. Water ejected above the atmosphere would turn into ice crystals in the vacuum of space.

The Ejecta Curtain

An impact excavates a crater by ejecting material in the shape of an inverted cone that expands with time. A meteorite with a diameter of two kilometers would complete the excavation phase in approximately thirty seconds. During this time, material would be ejected at ballistic velocities. The innermost ejecta are launched first and travel fastest in parabolic trajectories. Ejecta originating further from the center are launched later and move more slowly, falling nearer the rim (Melosh 1989). Computer models generally assume ejection angles close to 45 degrees. However, the width-to-length ratios of the Carolina Bays indicate impact angles averaging approximately 35 degrees, which would also correspond to the launch angles. The explanation for the difference in angles may be that the impact on the ice sheet created a rapidly expanding plume of steam that broadened the shape of the conical ejecta curtain, thus creating the lower ejection angles. A comprehensive survey of the Carolina Bays would be necessary to obtain a more precise average of the impact angles.

Even though the number of Carolina Bays is estimated to be approximately 500,000, there must have been millions of ice boulders with different speeds and trajectories in the ejecta curtain. The swarm of so many flying ice boulders would have had mid-air collisions that changed trajectories and formed some heart-shaped bays (Prouty, 1952). The bays resulting from diverted projectiles would not be aligned toward the point of origin, but the largest and most massive objects would resist changes to their trajectories and create bays more accurately oriented toward the point of the extraterrestrial impact. The determination of the point of origin should assign greater importance to the axial orientation of large well-defined bays.

Time of Emplacement

There are two factors that determined the time of formation of the bays. The first one was the initial launch time, which would have a maximum variance of about thirty seconds between the first and last ice boulders ejected during the excavation phase of the extraterrestrial impact. The second factor affecting time of emplacement was the launch angle; ice boulders launched at steep angles would have longer flight time than those launched at lower angles with the same range. The variation in the width-to-length ratios of the Carolina Bays is due to the different impact angles at which the bays were created. The variation in the sample presented earlier, corresponds to angles between 33 and 39 degrees. This means that the ejecta curtain of the extraterrestrial impact had an angular dispersion of about 6 degrees.

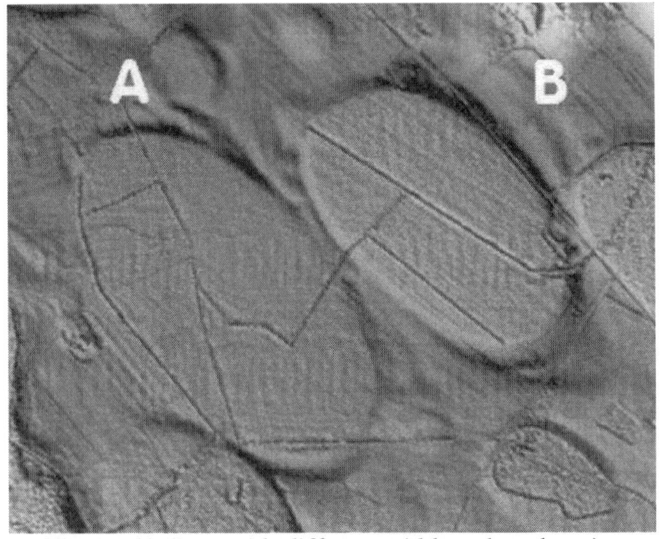

Figure 40. Bays with different width-to-length ratios (Lat. 34.841, Lon. -79.221)

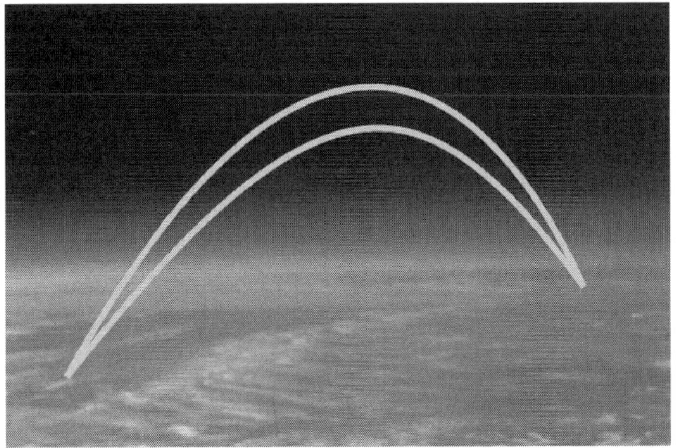

Figure 41. Trajectories with the same range at different launch angles

The sequence of emplacement can be determined by the principle of superposition for Carolina Bays that overlap. Older bays are overlaid by newer bays. If the bays were created by impacts, all of them were emplaced within ten minutes of each other; the term "older" could mean just a few seconds older.

In the case where neighboring bays do not overlap, their sequence of emplacement may be estimated from their flight time based on the width-to-length ratios. **Figure 40** shows adjacent bays with different width-to-length ratios, and **Figure 41** illustrates trajectories with the same range at different launch angles.

The bay marked **A** is more elongated than the bay marked **B**. Bay **A** has axes of 637×1166 meters with a ratio of 0.546, which corresponds to an impact angle of 33.114°. Bay **B** has axes of 543×942 meters with a ratio of 0.576, which corresponds to an impact angle of 35.198°. Assuming that the launch angles of the ice projectiles are the same as the impact angles calculated from the bays, the ballistic equations indicate that for projectiles traveling a distance of 1120 km, the projectile corresponding to **A** with the smaller launch angle and lower trajectory had a speed of 3.465 km/sec and a flight time of 386 seconds. The projectile for **B** with the higher trajectory had a speed of 3.415 km/sec and a flight time of 401 seconds. Thus, the projectile for **A** would have reached its destination 15 seconds before the projectile for **B**. This determination would be impossible to make using geological data because the impacts were formed essentially contemporaneously.

The Extinction Event

The distribution of the Carolina Bays and Nebraska Rainwater Basins indicates that chunks of glacier ice fell within a radius of 1500 kilometers from the point of an extraterrestrial impact in the Great Lakes region. The area covered by the ejecta extends from the Rocky Mountains to the eastern coast of the United States.

Firestone (2007, 2009) proposed that airbursts and fires from an extraterrestrial impact could have caused the megafauna extinction and the demise of the Clovis culture. However, the density of the Carolina Bays suggests that the extinction event was caused by a horrific pelting from the ice ejected by the extraterrestrial impact. Examine the following image of the area around Tatum, South Carolina (**Figure 42**). The saturation bombing by the huge ice boulders completely obliterated the landscape. Ask yourself the question: Where could I have stood to avoid being hit by an ice boulder? There was no safe place. Each bay corresponds to an impact with energy of 13 kilotons to 3 megatons of TNT explosive power.

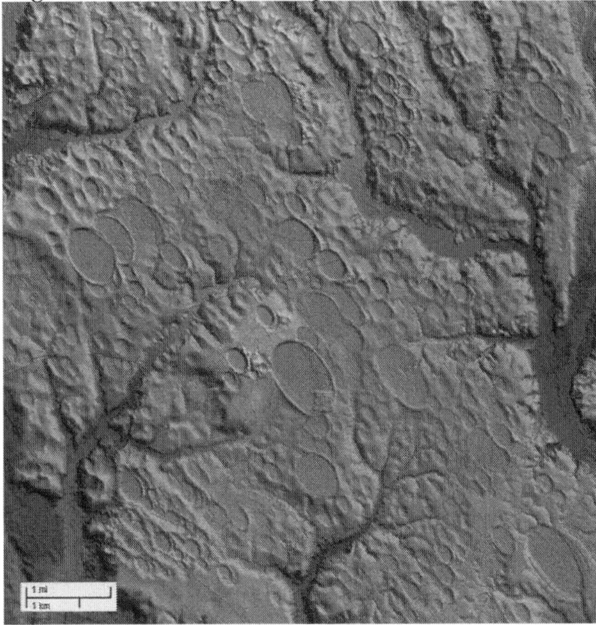

Figure 42. Carolina Bays near Tatum, South Carolina (Lat. 34.6513, Lon. -79.5850)

Many scientists worry that human civilization could be endangered if the Earth is hit by a meteorite in the future. The Carolina Bays suggest that this may already have happened in our not so distant past. The extinction of the Clovis people in North America may have changed the course of history and allowed human culture in Eurasia to take the lead.

Timeline of the Extraterrestrial Event

The ballistic equations and the clues provided by the Carolina Bays make it possible to reconstruct the timeline of the events at the time of the impact.

At T-minus-zero, an extraterrestrial object of 2 to 3 kilometers in diameter impacted the Laurentide ice sheet near Michigan. The impact fractured the ice sheet and melted and vaporized large quantities of ice. Ice chunks and liquid water were ejected propelled by clouds of steam under great pressure.

Within two seconds after the impact, seismic shock waves radiated at 5 km/sec from ground zero.

Twenty seconds after the impact, the water and glacier ice boulders ejected at 3 to 4 km/sec rose above the atmosphere.

At T-plus 30 seconds, the water ejected above the atmosphere or carried by the ice boulders turned into ice crystals that went into low Earth orbit.

Four minutes after the extraterrestrial impact, the seismic shock waves from the primary impact reached Nebraska and the Carolina coast liquefying the sandy terrain near the water table.

Seven minutes after the impact, the ice boulders reentered the atmosphere and started crashing into the ground accompanied by sonic booms. The ice boulders penetrated the liquefied ground creating conical cavities.

The high energy of the impacts shook the ground and contributed to the liquefaction of the soil and the viscous relaxation of the conical cavities. Ice striking solid ground shattered upon impact covering the ground with ice pieces. The saturation bombing by the huge ice boulders killed fauna and destroyed their habitat.

Ten minutes after the extraterrestrial impact, the ice bombardment stopped. The solid ground was covered with half a meter of ice pieces and the liquefied ground was covered with conical cavities that had transformed into elliptical bays by the vibrations of the bombardment.

Twenty minutes after the extraterrestrial impact, all was quiet again as the sonic booms became inaudible. The ice crystals in low Earth orbit diminished the light of the Sun and the ice pieces that covered the ground increased the albedo of the Earth, triggering a cold event.

The sound of the cosmic explosion by the Great Lakes traveled at 340 meters per second and reached the East Coast approximately one hour after the impact, but by that time there was no one alive to hear it.

Although the events proposed by the Glacier Ice Impact Hypothesis could have occurred at any time during the ice ages in North America, the late Pleistocene, 12,900 years ago, provides the best fit. This is the time of the onset of the Younger Dryas cooling event, the time of the extinction of the North American megafauna, and the disappearance of the Clovis culture.

* * *

SOIL LIQUEFACTION

The second premise of the Glacier Ice Impact Hypothesis is that the unconsolidated ground in the target areas was liquefied by the seismic shock waves from the primary and secondary impacts. The reason for this requirement is that impacts by ice projectiles can only make conical cavities on liquefied or viscous ground. If the ground is hard, the brittle ice projectiles will shatter without making conical cavities. Establishing that the terrain on which the Carolina Bays are found could have been liquefied for the formation of conical cavities strengthens the plausibility of the hypothesis.

One of the characteristics of the Carolina Bays and Nebraska Rainwater Basins is that they occur only in unconsolidated soil. There are no bays on hard ground. Saturated unconsolidated soil, such as that on which the bays are located, is easily liquefied by seismic shock waves. Soil liquefaction or acoustic fluidization is a phenomenon in which a saturated soil becomes like quicksand and behaves like a viscous liquid in response to the applied stress of soil vibrations.

Sandy soil can be liquefied by vibrations when there is water near the surface. The vibrations suspend the sand grains in the water and reduce friction causing the soil to flow like a liquid. The 1964 earthquake in Niigata, Japan, which had a magnitude of 7.5 liquefied the ground and caused some buildings to sink and tilt (**Figure 43**).

The earthquake in Christchurch, New Zealand in February 2011 had a magnitude of 6.3 and produced significant liquefaction. The vibration of the earth turned sandy soils into quicksand that swallowed cars and displaced roads (**Figure 44**). Even smaller aftershocks measuring 5.7 liquefied the soil. In general, seismic events of magnitude 6.0 cause sufficient vibration to liquefy saturated soil.

Figure 43. Buildings toppled when the soil liquefied during the 1964 earthquake in Niigata, Japan

Figure 44. Vehicles submerged in liquefied soil during the 2011 earthquake in Christchurch, New Zealand

Terrain of the Carolina Bays

A report from the U.S. Geological Survey by Eimers (2001) shows that the areas of North Carolina that contain Carolina Bays have the water table within a few meters from the surface. These areas would be susceptible to liquefaction today and probably also at the time that the bays formed.

Figure 45 from the U.S. Geological Survey shows the estimated depth to water in North Carolina. The light blue color along the coastline indicates a water table within 5 feet, or 1.5 meters, from the surface. This coincides exactly with the areas that have Carolina Bays.

Terrain of the Nebraska Rainwater Basins

In Nebraska, the bays also are found in soil that is close to the water table. **Figure 46** from the University of Nebraska at Lincoln shows the depth to water in several shades of blue. Many bays can be found in the area with the light blue color where the water table is close to the surface in the flood plain of the Platte River, particularly in Phelps and Kearney counties.

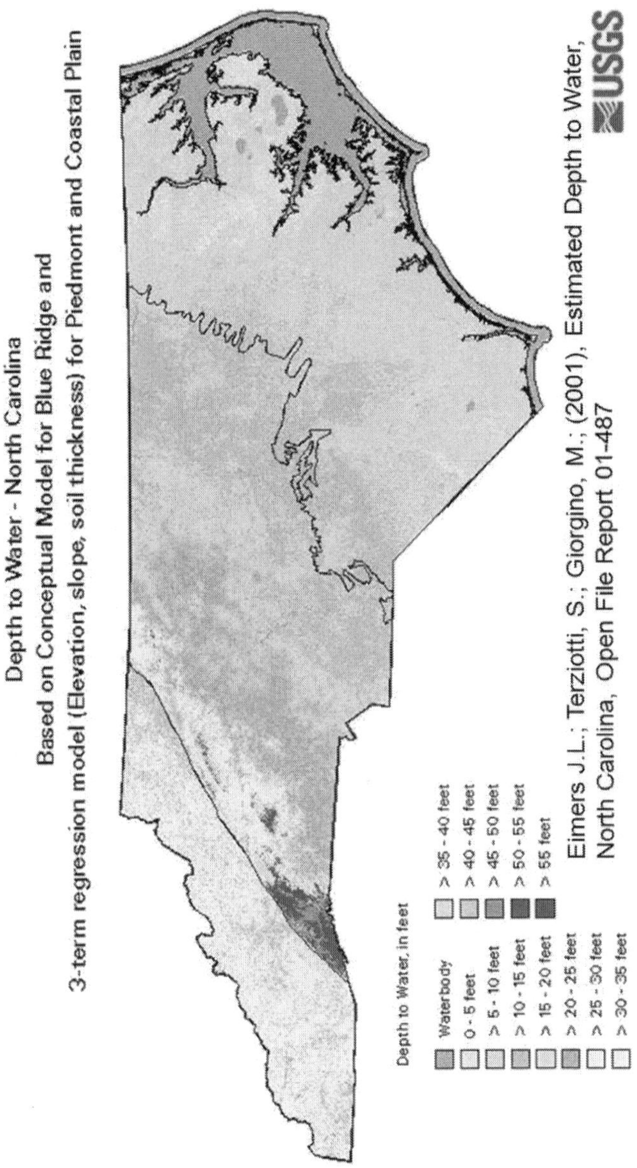

Figure 45. Depth to water in North Carolina

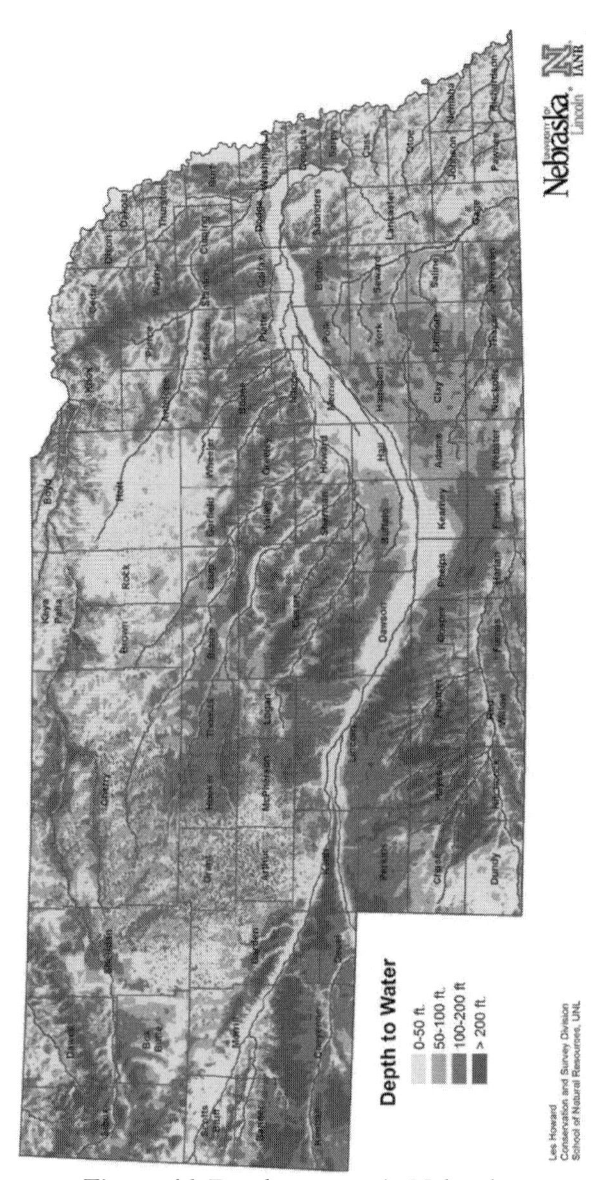

Figure 46. Depth to water in Nebraska

Sources of seismic shock waves

The fact that the Carolina Bays in the East Coast and in Nebraska occur in soil that is close to the water table and has the potential of being liquefied by seismic shocks suggests that this characteristic was important in the formation of the bays.

Seismic shock waves travel with a velocity ranging from approximately 2 to 8 km/sec. Primary compression waves, called P-waves, travel at 5 km/s in rocks such as granite. The seismic waves generated by the extraterrestrial impact in the Great Lakes region would have taken from 3 to 5 minutes to reach the coastal areas of the United States. Even though the shock of the primary impact would have been felt 1200 kilometers away, it might not have had enough energy to fluidize the ground. However, the impacts of the ice boulders arriving 3 to 6 minutes later definitely would have liquefied the surface.

The bays with a diameter of one kilometer were made by impacts with an energy of about 3 megatons, which is equivalent to an earthquake of 7.54 magnitude. The smaller bays with diameters of 220 meters were made by impacts with an energy of 13 kilotons of TNT, which is about the same energy as a 6.0 magnitude earthquake. Taking into consideration the massive bombardment of the landscape, there is no doubt that all the unconsolidated terrain in the target areas that was close to the water table became liquefied and provided a suitable surface for the creation of conical impact cavities.

The liquefaction of the surface may also have played a role in the extinction event because any land animals that were not hit by the glacier ice chunks would have been trapped in the resulting quagmire.

* * *

IMPACTS ON SOLID AND VISCOUS SURFACES

The discussion about the ellipticity of the Carolina Bays proposed that the bays originated as oblique conical cavities that were remodeled later by geologic processes into elliptical bays. The third premise of the Glacier Ice Impact Hypothesis is that oblique impacts of glacier ice boulders on the liquefied ground created slanted conical craters. In this section, we will examine impacts at different speeds on different types of surfaces to determine the conditions under which conical cavities can be created.

An **impact** is the action of one object coming forcibly into contact with another or the collision of two bodies at a non-zero velocity. It is important to distinguish between ballistic impacts and extraterrestrial impacts because they have different effects on their target surfaces.

A **ballistic impact** is a collision at a velocity not exceeding the escape velocity of a moon or planet. For the Earth, the maximum speed for a ballistic impact would be approximately 11 km/sec. Objects hurled from the Earth at greater velocities could leave the Earth and never return.

An **extraterrestrial impact** is a collision between the Earth and an object from outside the Earth, such as an asteroid or a comet. Asteroids that orbit the Sun between Mars and Jupiter can strike the Earth at velocities of 17 km/sec, whereas comets that originate in the outer Solar System in the Oort Cloud can hit the Earth at speeds of around 51 km/sec. The further out that a body is from the Sun, the greater velocity it will acquire from the Sun's gravitational attraction by the time it reaches the orbit of the Earth. In order to make a crater, a celestial object has to survive its passage through the atmosphere. Many extraterrestrial objects just burn up in the atmosphere and never impact the Earth.

Figure 47. Meteor Crater with a diameter of 1.2 km formed from the impact of an iron meteorite. (Lat. 35.029187, Lon. -111.025605)

Meteor Crater in Arizona, also known as Barringer Crater, was the first recognized impact crater on Earth (**Figure 47**). American geologist Eugene M. Shoemaker studied extensively the structure of Meteor Crater (Shoemaker 1960). He established the field of planetary science through the careful observation of large impact structures and microscopic examination of the minerals that could be used as indicators of a crater's origin. The methodology used by Shoemaker has enabled scientists to distinguish volcanic craters from extraterrestrial impact craters, even after the craters have had significant structural alteration by wind or water erosion. Cores taken from the bottom of the crater can be used to provide information about the layers underlying the cavity. The material surrounding the crater may indicate the types of forces that shaped the crater. Microscopic examination of the minerals in the crater can be used to look for crystals deformed by intense pressure. In his analysis of Meteor

Crater, Shoemaker found shocked quartz (coesite), a form of quartz with a crystal structure that is only produced at very high pressure; he had encountered similar crystals at the Nevada test sites where atomic bombs had exploded.

An extraterrestrial impact on a solid surface goes through stages of compression, excavation and modification (Melosh, 1989). During the **compression stage**, the projectile is destroyed and most of its kinetic energy is transferred to the point of impact. The energy vaporizes, melts and fluidizes the target material. In the **excavation stage**, a hemispherical shock wave expands from the point of impact and creates a bowl-shaped crater. At the hyperspeeds of extraterrestrial impacts, which vary from 14 km/sec for asteroids to 51 km/sec for comets, the transfer of kinetic energy during the compression stage is virtually instantaneous and even moderately oblique impacts create hemispherical shock waves and bowl-shaped craters. The **modification stage** is controlled mainly by gravitational forces. The ejected material falls to the surface creating ejecta rays and inverted flaps. Molten minerals may collect at the bottom of the crater and combine with broken rock fragments to form a breccia lens. The crater walls may become unstable and slide down into the cavity to form the final configuration of the crater.

The energy of an extraterrestrial impact is so great that the meteorite usually vaporizes completely by the time it has buried itself in the target surface. Once the projectile is gone, its energy is transmitted as a hemispherical shock wave that creates a similarly shaped crater. When viewed from above, most impact craters are circular.

The recognized and expected markers of an extraterrestrial impact consist of raised rims around the crater (**Figure 48**), meteorite fragments within or surrounding the crater, petrographic shock indicators such as crystals with planar deformation features (PDFs), layers of fractured rocks (breccias) within the crater, and shatter cones in the surrounding rocks which are aggregates of mineral crystals fused together in conical shapes by the passage of an intense shock wave (French 1998).

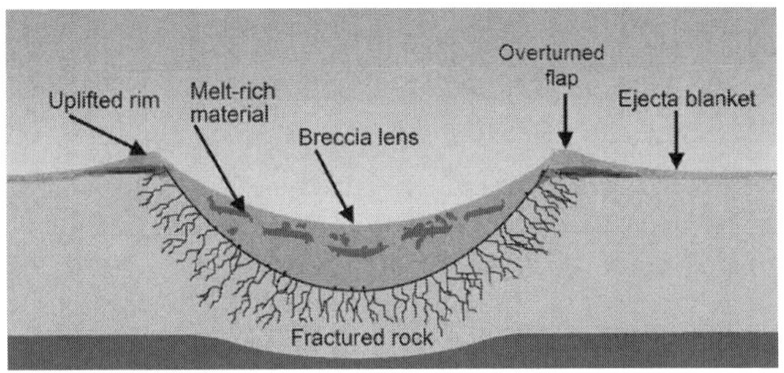

Figure 48. Typical crater from a cosmic impact (modified from French)

The raised rims result from the uplift created by lateral displacement of material as the crater is formed, and from material that is excavated and folded over the rim of the crater forming an overturned flap. When meteorite fragments cannot be found in or around the site, the proposed impact site should at least contain traces of siderophile elements, such as iridium, that are more abundant in space rocks than in the Earth. The formation of petrographic shock indicators, like shatter cones and planar deformation features, require minimum pressures of 5 to 30 Gigapascals (Melosh 1989).

Impacts on viscous media

The term "viscosity" is a measure of the resistance to deformation of a medium under stress. The term is usually applied only to gases and liquids, although it may also be applied to solids that deform without breaking.

We have seen that hyperspeed impacts destroy the projectile during the compression stage when the projectile makes contact with the target and then creates a hemispherical shock wave. By contrast, the impact of a projectile on a viscous medium at speeds of less than 4 km/sec usually does not destroy the projectile. The excavation stage starts as the projectile travels through the medium creating a conical shock wave until it is stopped by friction.

The crater created by an impact is determined by the type of shock wave generated by the projectile. Hemispherical shock waves create bowl-shaped craters; conical shock waves create conical craters whose stability depends on the composition of the target medium. Conical impact cavities are ubiquitous, but they are seldom noticed because they disappear very quickly. Conical cavities created by impacts in fluids and gels disappear in fractions of a second, but a viscous surface with low elasticity will retain the conical shape, and the conical cavity will gradually be modified by gravity through viscous relaxation.

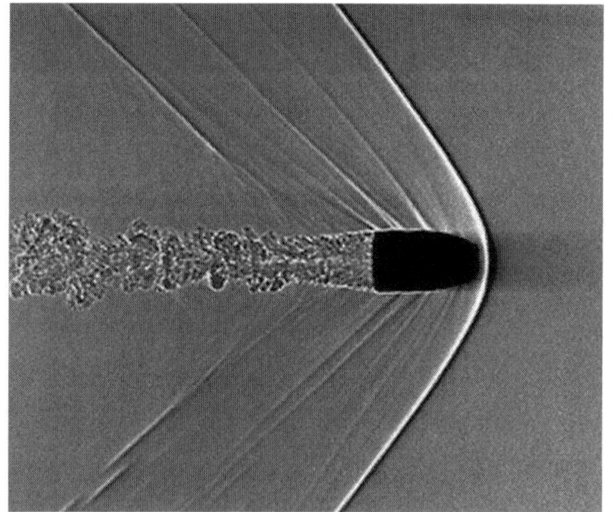

Figure 49. Conical shock wave of a bullet traveling in air

The trajectory of a bullet in a gas creates a shock wave that is usually considered conical, but it is actually a paraboloid (**Figure 49**). The gaseous medium is not able to retain the shape of the conical shock wave because of its low density and fluidity.

Impacts on viscous elastic targets

Ammunition is frequently tested on ballistic gelatin targets that are viscous but also elastic. Ballistic gelatin is made by dissolving gelatin powder in water so that it closely resembles the density and viscosity of human and animal muscle tissue. Hollow point bullets that shatter on impact create complex trails in the gelatin. Bullets that retain their integrity form conical cavities that quickly collapse due to the elasticity of the gelatin. High-speed photographs of bullet trajectories through gelatin often reveal irregular cavitation caused by the spin or tumbling of the bullets.

Target materials that are more dense than a gas, such as gelatin or water, show conical cavities during the transit of an impacting projectile (**Figures 50** and **51**). Gelatin quickly regains its original shape due to its elastic nature and only a linear track of the projectile remains. Water flows under the influence of gravity to fill the conical cavity leaving no trace of the impact.

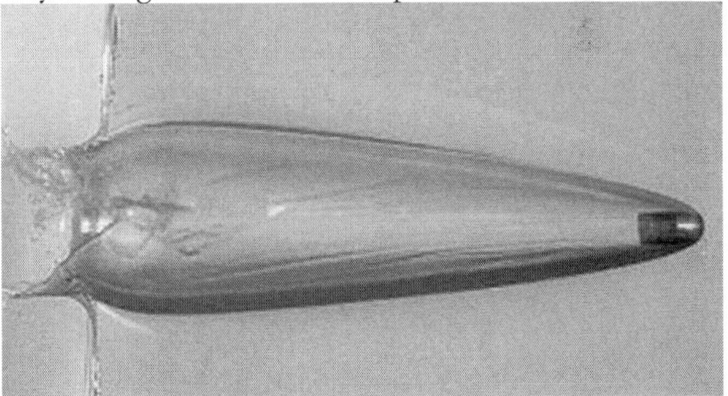

Figure 50. Conical cavity in ballistic gelatin

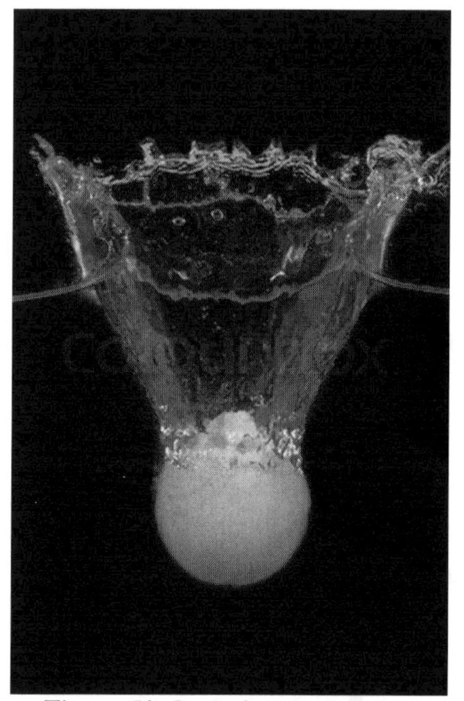

Figure 51. Conical cavity in water

Small impacts on water illustrate the elasticity of the surface and the effect of surface tension in a medium of low viscosity. The elasticity is due in part to the force of gravity, which tends to create a level surface. When a projectile penetrates the water surface, it creates a conical cavity. The impact generates a shock wave and the displaced water forms a crown-shaped ejecta curtain (**Figure 52**). Once the energy of the projectile has been dissipated, the cavity becomes a region of low pressure. The centripetal rush of water collides at the center of the cavity and forms a rising column of water (**Figure 53**). When the column collapses back onto the surface, a secondary cavity is formed which repeats the process, each time with less intensity. The oscillation forms concentric ripples typical of elastic deformation of Newtonian fluids under the force of gravity. Small spherical droplets form at the extremities of the ejecta curtain where surface tension is the dominant force.

Figure 52. Impacts on water create a crown-shaped ejecta curtain

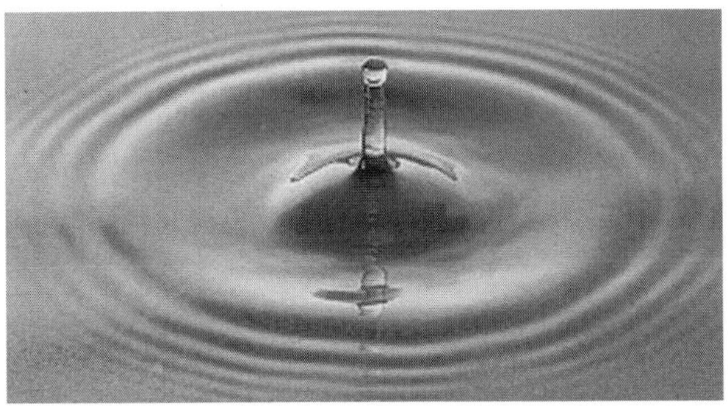

Figure 53. A rising column of water forms when water flows into the impact cavity

Impacts on viscous surfaces with low elasticity

When the density of viscous target material increases to a plastic consistency of low elasticity, the conical shape of the transient cavity is retained for a significant period of time. The impact of a bullet on modeling clay (**Figure 54**)

or of an ice ball on a sand-clay mixture (**Figure 59**) creates conical cavities that persist for a relatively long time and modify slowly under the force of gravity. As the density of the target material increases further toward a solid, the projectiles disintegrate on impact and transfer all their kinetic energy to the target thus creating hemispherical shock waves that produce the typical bowl-shaped craters. Conical cavities can only be made when the projectile is able to travel through the medium without disintegrating.

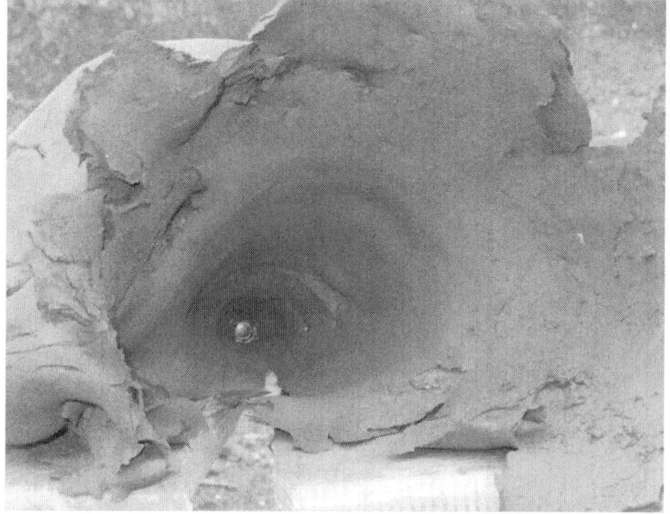

Figure 54. Conical cavity created by a bullet on modeling clay

Books on impact cratering generally do not discuss conical impact cavities. The omission occurs also in the geology and astronomy literature. One exception was the report by Garvin (2011) about the Linne crater on the Moon that was discovered to have a conical structure based on observations made by the Lunar Reconnaissance Orbiter. Conical impact cavities receive little attention because they are generally transient and change rapidly.

Experimental conical craters

Figures 55 through **58** illustrate the formation of a conical cavity by the impact of a marble fired with a slingshot at a viscous surface consisting of equal parts of pottery clay and sand with enough water to have the consistency of mortar. The surface of the target was thinly covered with colored sand to enhance analysis of the images. This non-Newtonian thixotropic sand-clay mixture has little elasticity and the material ejected by an impact is not influenced substantially by surface tension.

Figure 55. Initial Surface

Figure 56. Start of Excavation

Impacts on viscous surfaces do not have a clear demarcation between the compression stage and excavation stage since the projectile is not destroyed. Upon impact, the projectile pushes target material out of its path, compressing it and accelerating it, while at the same time the projectile is slowed down by the target's resistance.

Figure 57. Maximum excavation

Figure 58. Final elastic modification

The penetration of the projectile causes displacement of material that forms an expanding ejecta curtain. In **Figure 57**, a portion of the projectile can be seen at the bottom of the conical cavity when the cavity reaches the point of maximum excavation. Under the influence of gravity, some of the ejected surface material follows ballistic trajectories while material at the edge of the cavity starts to form an overturned flap. The start of the modification stage consists of an elastic rebound (**Figure 58**) that decreases the size of the cavity and buries the projectile.

The final elastic modification occurs while the last of the ejected material is still in flight approximately one crater diameter from the point of impact. The resulting conical cavity becomes relatively stable after all the kinetic energy of the impact has been dissipated. At this point, gravity becomes the major force acting on the cavity, and modification continues more slowly through viscous relaxation.

Figure 59. An impact by an ice ball on a viscous surface creates a conical cavity with an overturned flap.

In the case where unconsolidated soil increases in density as a function of depth from natural stratification, a projectile impacting at an oblique angle will encounter increasing resistance traveling deeper compared to traveling horizontally. The asymmetric resistance gradient will change the trajectory of the projectile into a path that creates a cavity shaped like a curved horn instead of a cone. This may affect the shape of the resulting bay.

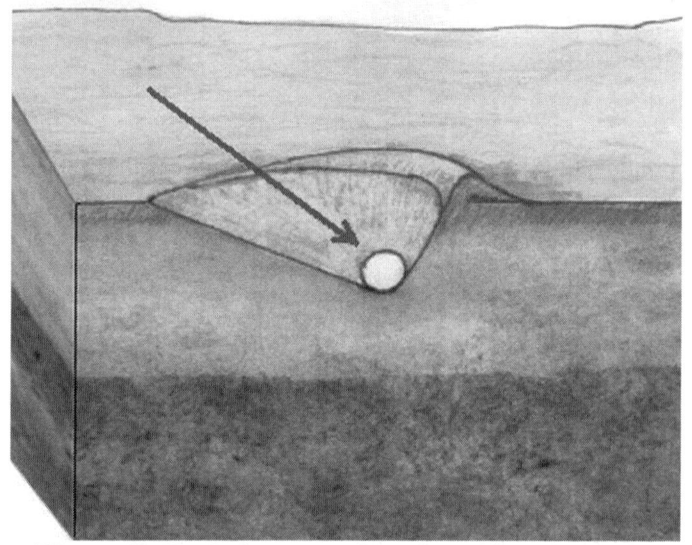

Figure 60. Vertical cross section of an oblique impact

Oblique conical cavities on liquefied ground are quickly remodeled into shallow elliptical depressions (**Figures 61** and **62**). The following images show the transformation of cavities made by ice on a clay-sand mixture. The modification of the conical cavities into elliptical bays by the action of gravity is discussed in more detail in the section about viscous relaxation.

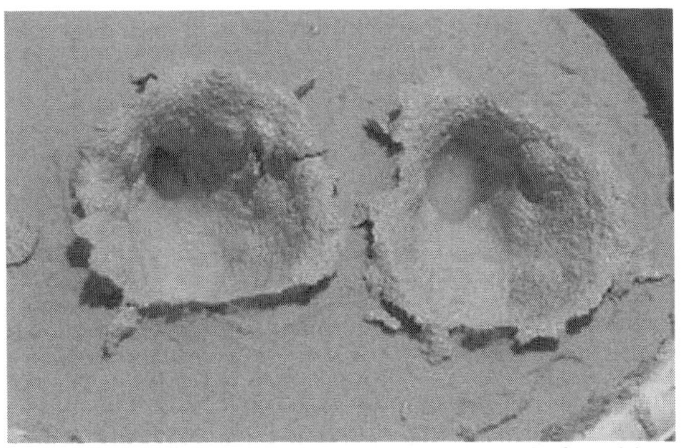

Figure 61. Oblique impacts create conical cavities

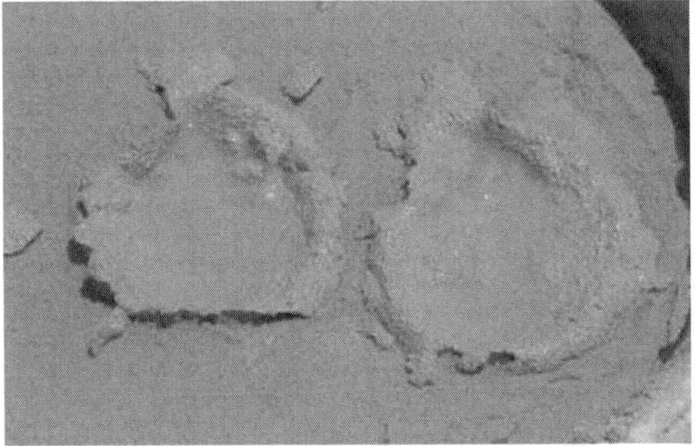

Figure 62. Conical cavities transform into bays

Overturned Flaps

Gamble et al. (1977) suggested that the raised rims around the Carolina Bays might be secondary rims formed over time from eolian sand deposits that buried a primary rim at the edge of an initial depression. However, raised rims are mandatory characteristics of impact craters. The raised rims around an impact cavity are formed during the excavation phase as the shock wave of the impacting body penetrates the ground, displaces material laterally producing structural uplift around the cavity and ejects material above the surface. The overturned flaps are created during the remodeling phase when the ejected material falls back to the surface under the influence of gravity (Maxwell, 1977). On a viscous target, the ejected material acts like a breaking wave in which the base of the wave meets resistance from the surface while the crest continues its forward motion. The final overturned flaps usually display inverted stratigraphy near the rim (Melosh 1989). **Figure 63** illustrates the development of an overturned flap.

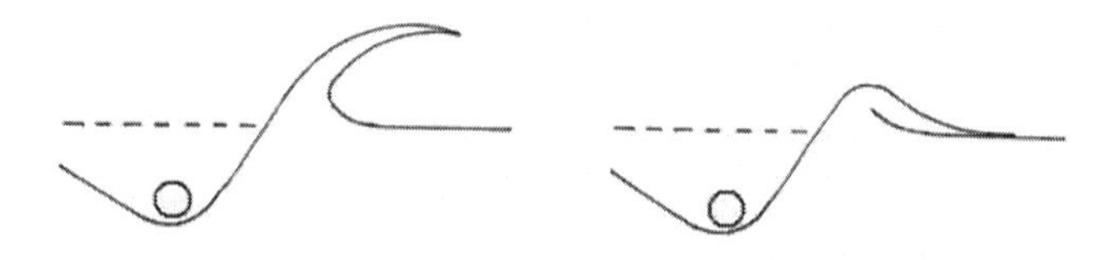

Figure 63. Formation of an overturned flap

Figure 64 indicates the locations of overturned flaps in several Carolina Bays. The southeast end of the bays has thicker rims because the oblique impacts pushed surface material in the direction in which the impacting projectile was traveling. This plowing action is also a characteristic observed in experimental impacts on a sand-clay medium. Finding evidence of inverted stratigraphy in the rims of the Carolina Bays would be a way to confirm the impact origin of the Carolina Bays independently of whether the site where the extraterrestrial impact occurred is found.

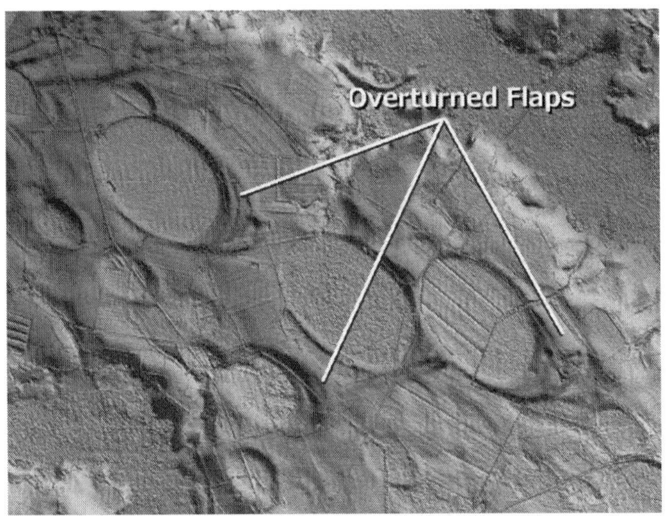

Figure 64. Examples of overturned flaps in the Carolina Bays. (Lat. 34.8572, Lon. -79.1832)

This section focused on the excavation phase of impacts on solid and viscous surfaces and presented a physical model that demonstrates that the overturned flaps of an oblique impact on a viscous surface form raised rims around a slanted conical cavity. An analogous impact mechanism on a larger scale could have produced the raised rims around the Carolina Bays.

* * *

VISCOUS RELAXATION

The fourth premise of the Glacier Ice Impact Hypothesis is that viscous relaxation reduced the depth of the conical cavities and remodeled them into elliptical bays. Hyperspeed impacts create different types of craters depending on the structure and properties of the target surface. Terrestrial craters with diameters of less than 4 km have a simple bowl shape, while larger craters typically have a central uplift from deeper strata or a multi-ring structure (French 1998). The uplift is created by the elastic properties of a solid surface that deforms and rebounds when impacted. By contrast, the Carolina Bays, even those with a major axis exceeding 10 km, have a uniform type of configuration due the way that impact cavities are formed on unconsolidated, saturated soils by projectiles at slower ballistic speeds.

Impacts have the immediate result of producing a cavity that is modified rapidly, but craters on most planetary bodies continue to change toward the limit of gravitational stability, which is a level plain. Materials that appear solid actually flow slowly under stress, and this flow increases as a function of temperature. Although the mechanics of crater collapse are not well understood, many of the properties of crater collapse can be described with a Bingham plastic model, which is a rheological model that incorporates plastic and viscous behavior (Melosh 1982, 1989).

Viscous relaxation is a plastic deformation process driven by gravity that tends to smooth out geological features by making hills less prominent and valleys less deep. The process is generally slow, but it can be speeded up by reducing the friction within the medium. For saturated sandy mixtures, this can usually be accomplished by vibrations that promote liquefaction, also called acoustic fluidization. The following images show the transformation of an experimental conical cavity on a sand-clay target.

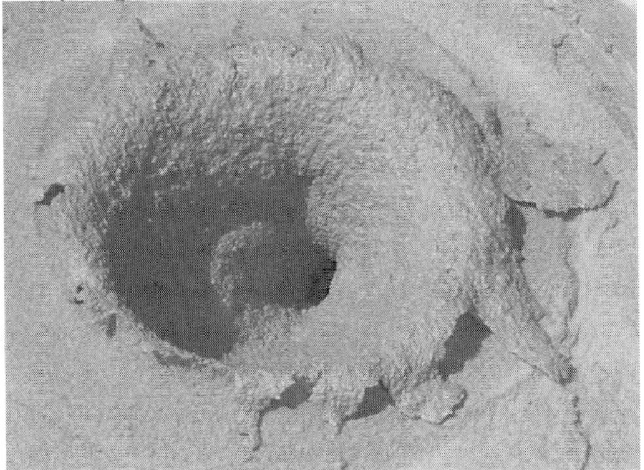

Figure 65. Initial conical cavity

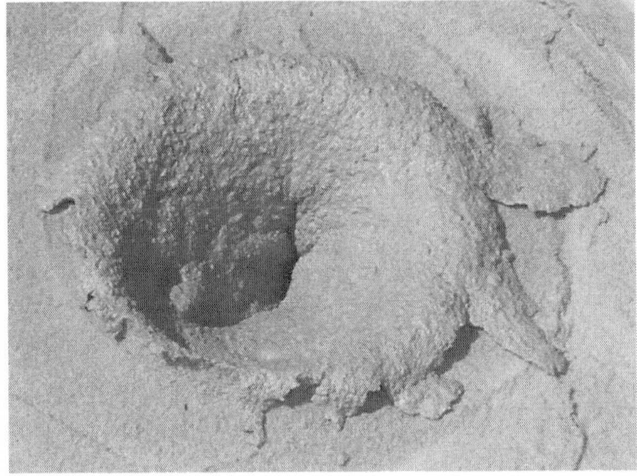

Figure 66. Start of viscous relaxation

Viscous relaxation reduces the depth of an impact cavity from the bottom up, essentially reversing the sequence in which the cavity was formed.

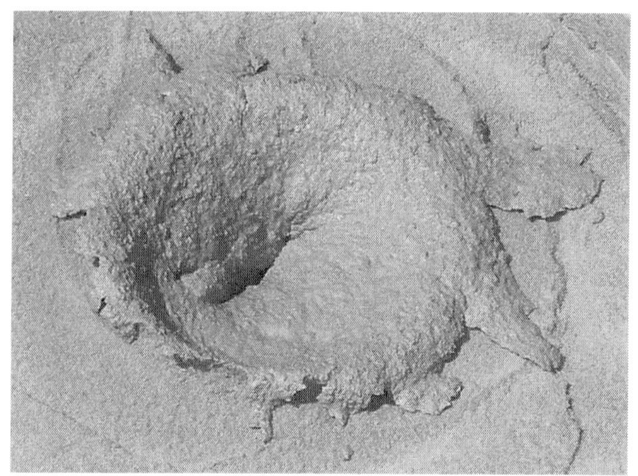

Figure 67. Continuation of modification

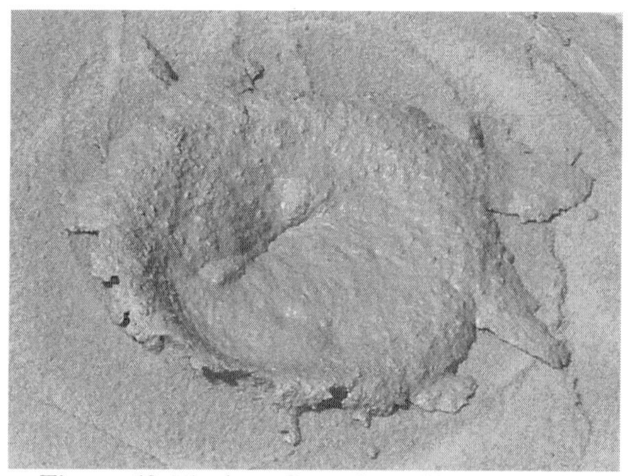

Figure 68. Final configuration of impact cavity

The modification of the cavity continues until the lateral centripetal flow of material cannot overcome the cohesion from internal friction. In the final configuration, a shallow depression is surrounded by an elevated rim made of the overturned flap created by the impact. Over time, water will seep into the cavity to create a pond. This is the mechanism proposed for the formation of the Carolina Bays.

Overlapping bays

None of the hypotheses of formation of the Carolina Bays by wind and water processes or by other geological processes, such as ground dissolution, have proposed a mechanism for creating overlapping bays with raised rims and specific width-to-length ratios. Impacts provide an answer to this problem because adjacent conical impacts are transformed by viscous relaxation into overlapping bays. The chronology of the impacts determines the manner in which the bays overlap, and the width-to-length ratio is simply dependent on the impact angle.

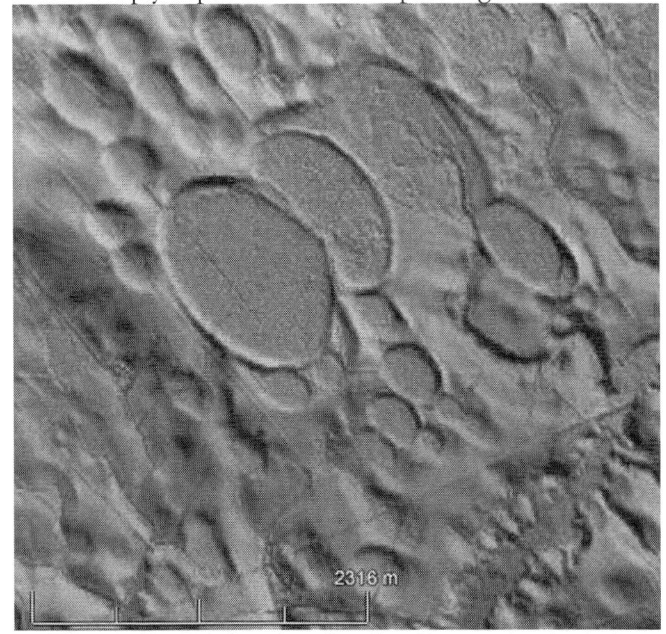

Figure 69. Overlapping Carolina Bays maintain their width-to-length ratios (Lat. 34.6445, Lon. -79.6362)

The following images illustrate the transformation of adjacent conical impact craters into overlapping depressions analogous to the Carolina Bays. Two adjacent impacts were made on a sand-clay target with ice ball projectiles fired with a slingshot. **Figure 70** shows the conical impact cavities.

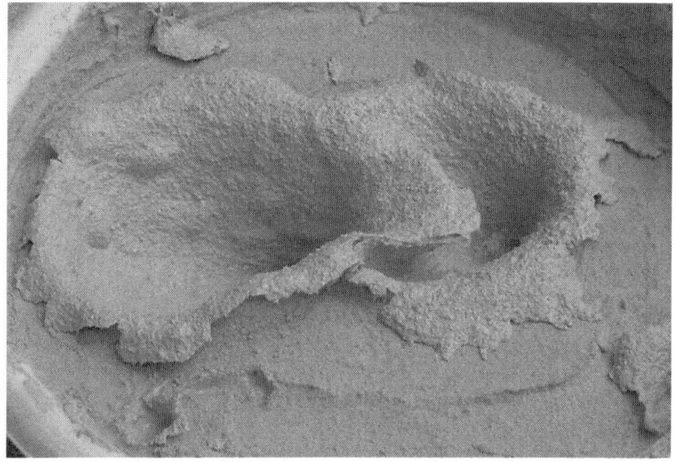

Figure 70. Adjacent conical impacts

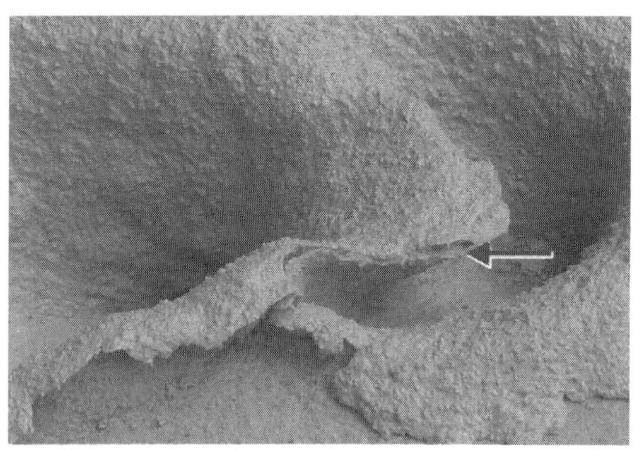

Figure 71. Boundary of inverted stratigraphy

Figure 71 is a detailed view showing a thin layer of colored sand in the overturned flap that marks the boundary of the inverted stratigraphy.

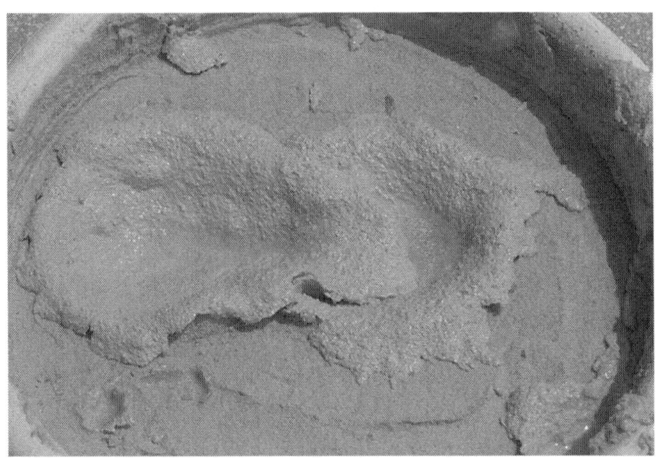

Figure 72. Transitional phase during viscous relaxation

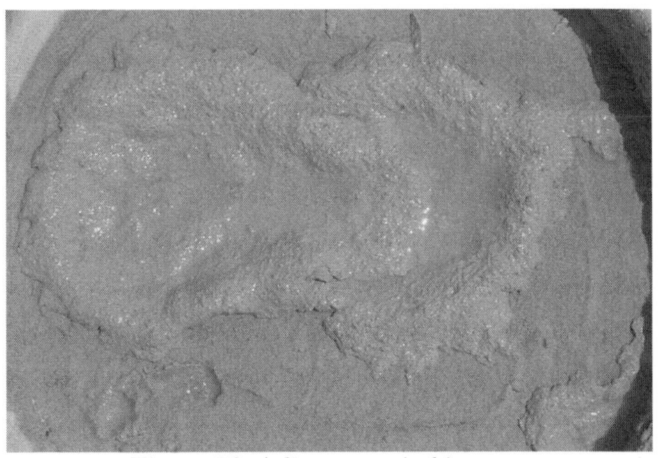

Figure 73. Adjacent conical impacts become overlapping bays

Figures 72 and **73** show the gradual reduction of crater depth due to viscous relaxation that creates the final depressions with raised rims. The overlap of the final depressions can be used to deduce the geochronology of the impacts because the most recent impacts always cover the preceding impacts as expected from the principle of superposition.

Stratigraphic Restoration

A feature of the Carolina Bays that has been used to argue that the bays are not impact structures is that the stratigraphy beneath the bays is not distorted as might be expected after an impact. The explanation for this is that the excavation phase of a ballistic impact on a viscous surface is not an explosive event like that produced by a hypervelocity impact. A projectile impacting a viscous surface penetrates the medium and parts it in a plastic deformation that can be partially reversed by viscous relaxation as shown in **Figure 74.**

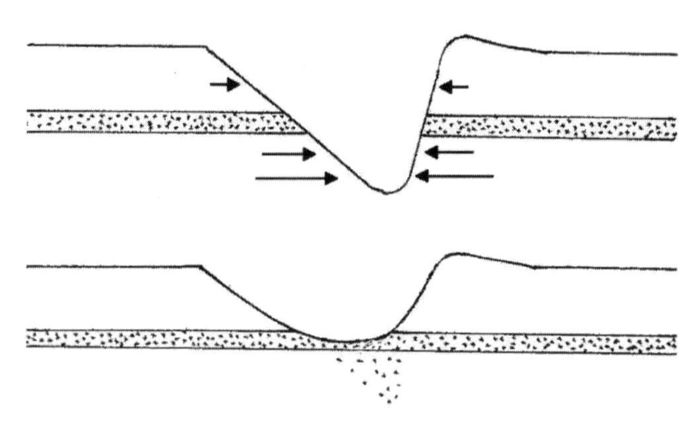

Figure 74. Stratigraphic restoration by viscous relaxation

A projectile at ballistic speed penetrates a viscous surface by parting the existing layers of the target during the excavation phase of the impact. All the energy of the projectile is spent moving the non-elastic material along its path. Viscous relaxation during the modification phase, which is driven by gravity, subdues topographic features, such as craters, and partially reconstitutes the stratigraphy by the flow of the material parted during the excavation. The leveling process gradually reduces the depth of the cavity, and ground vibrations may accelerate the rate of viscous relaxation by reducing the shear resistance of the soil.

A cavity in a viscous surface is filled by flow of the material surrounding the deepest part of the cavity. Pressure increases with depth and this creates a velocity gradient that promotes faster centripetal lateral flow at the bottom of a cavity in a uniformly viscous medium. The increase of pressure **P** with depth is given by the equation $\mathbf{P} = \rho \mathbf{g} \mathbf{h}$ where ρ is the density, **g** is the gravitational constant and **h** is the depth.

The cavity cannot be filled by upwelling of substrata because the pressure from the weight of the terrain around the cavity, which causes the remodeling, cannot be fully transmitted to the substrata as long as material is being diverted by lateral laminar flow to fill the cavity. The flow into the cavity stops when the pressure is insufficient to overcome the frictional forces of the medium. Viscous relaxation reverses the sequence in which the cavity was created. This is somewhat different from the collapse of a conical cavity made in ballistic gel by the passage of a bullet because once the bullet stops, the gel that was parted by the bullet comes back together again, but through elastic forces, rather than the force of gravity. The following images illustrate that the material that fills any level is the same material that was parted by the shock wave of the impact.

Stratigraphic Restoration Experiment

The following images show the crater made by the oblique impact of an ice ball on a sand-clay target that had been prepared with an underlying red layer approximately two centimeters below the surface. The penetration of the projectile through the red layer dragged along some of the red material. **Figures 76** through **78** show the gradual reduction in depth as the viscous material adjacent to the deepest part flows to fill the cavity. The red layer remains at the same level and disappears from view when the impact structure reaches its final configuration because the cavity is filled by centripetal lateral flow of material. If the cavity had been filled by upwelling of material from lower strata, the red layer would have been raised and exposed to the surface.

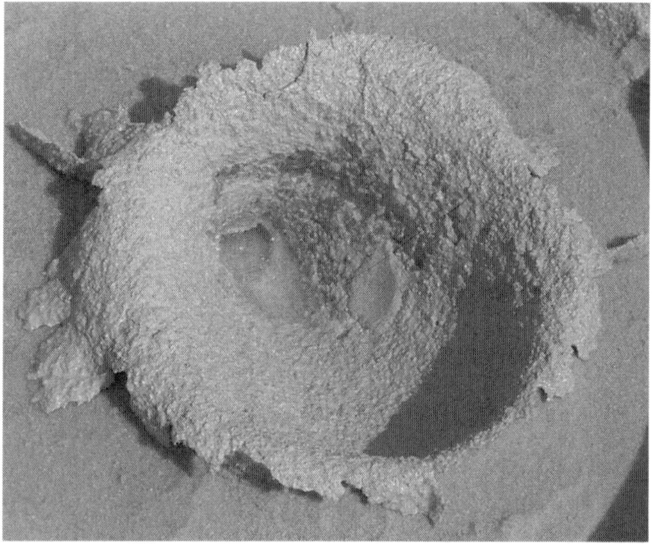

Figure 75. Ice projectile penetrates through a colored layer

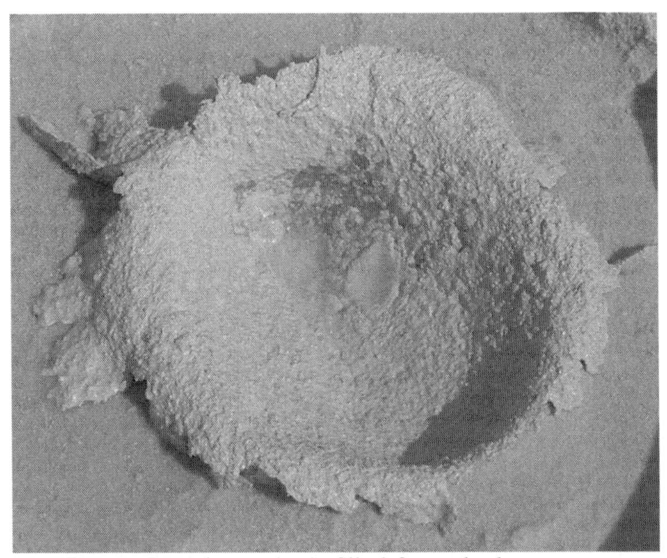

Figure 76. The cavity is filled from the bottom up

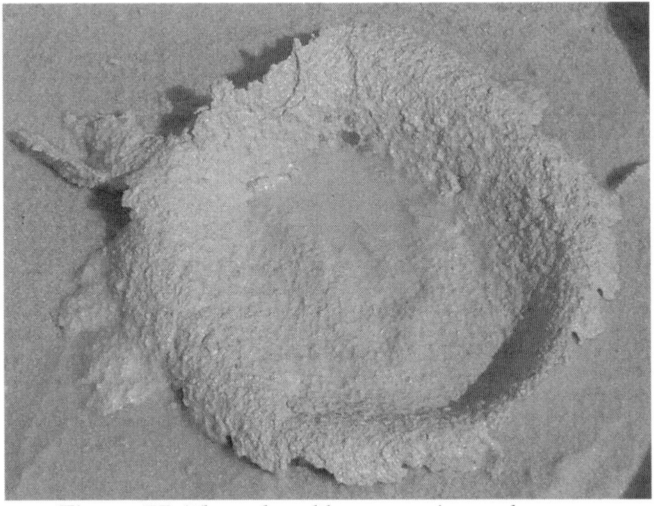

Figure 77. The colored layer remains at the same level as the cavity is filled

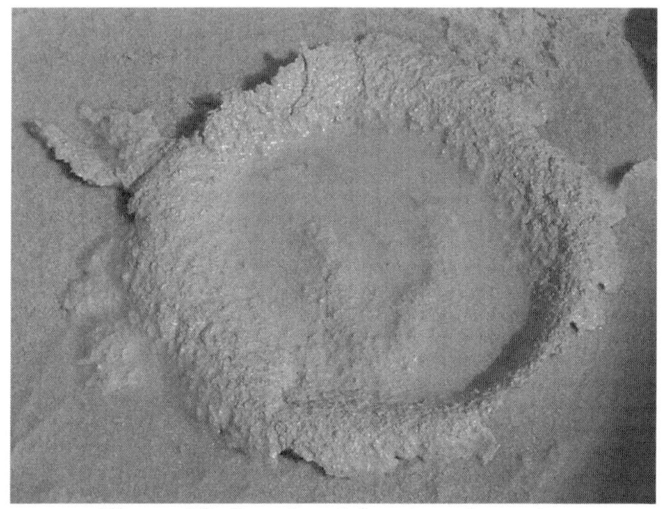

Figure 78. Centripetal flow into the cavity from the bottom up restores the stratigraphy

Vibrations speed up the viscous relaxation of the saturated clay-sand surface. During the creation of the Carolina Bays, adjacent impacts would have provided the vibrations necessary to reduce the depth of the conical cavities.

* * *

DATING THE CAROLINA BAYS

There are several reasons why the Carolina Bays have not been generally accepted as impact structures. One of them is that the bays do not show signs of shock metamorphism or traces of siderophile elements that would be expected for hyperspeed extraterrestrial impacts. This objection can be overcome by noting that if the bays were created by ballistic impacts of terrestrial glacier ice, the low speed impacts would not produce high pressures or carry unusual chemical markers to create those signatures.

The main objection to an impact origin of the Carolina Bays comes from the studies showing that the terrains on which the bays are found have a wide variety of dates, some older than 100,000 years. Since very accurate dating methods have been used, it is easy to conclude that all the Carolina Bays could not have been formed at the same time by an impact event.

The only way of overcoming this objection is to show that the dates measure the age of the terrain and not the time at which the bays were formed. Dissociating the date of the material from the date of the structures would make it possible to consider that all the bays were created by a single event. This is analogous to saying that the date of construction of a house made of stones should not be determined by measuring the geologic age of the stones. The discussion below presents some reasons why the dates of the terrain may not correspond to the dates of bay formation.

The Geologic Dates

Many papers have been written about the dates of the Carolina Bays. Rather than mention them individually, we will focus on the paper by Brooks (2010) which provides references to the work of other authors and describes recent findings. According to Brooks, Optically Stimulated Luminescence (OSL) dating and the deep sea and ice core oxygen isotope records, have greatly improved the dating and interpretation of sparse, discontinuous terrestrial records of Late Pleistocene conditions.

OSL dating estimates the time since last exposure to sunlight for quartz sand and similar materials with a range of 150,000 years before the present. OSL technology takes advantage of the fact that cosmic rays and ionizing radiation from naturally occurring radioactive elements in the earth can cause electrons to become trapped in the crystal structures of buried quartz and other minerals. OSL is able to free the trapped electrons and produce luminescence in proportion to how long the quartz has been buried. Exposure of quartz to sunlight eliminates the trapped electrons and resets the clock of the luminescence signal. Samples for OSL dating have to be taken in subsoil not exposed to light and they are processed in darkness.

The OSL dates are interpreted by correlation to the oxygen isotope records of global climate change. In this way, it was determined that the rims of the Carolina Bays were formed during and just prior to glacial stadials. Where concentric rims occur, the rims are progressively younger toward the center of the bay. This is taken as evidence that the bays are not single-event features, but that they evolve episodically over a long period of time.

What is wrong with the dates?

The dates of formation of the Carolina Bays cannot be discussed separately from the hypotheses of their formation. The water-and-wind hypotheses require a protracted period of time estimated to be from 30,000 to 140,000 BP, whereas an impact hypothesis requires the contemporaneous formation of all the bays at any time during the Pleistocene Epoch between 11,700 and 260,000 years ago when North America had a thick cover of glacial ice.

If the wind and water hypothesis had a verifiable explanation for the formation of perfectly elliptical bays and for their radial orientation, then it might be possible to believe that the geologic dates were meaningful. However, there is no supporting evidence that wind blowing over water ponded in shallow depressions can create perfectly elliptical bays oriented toward the Great Lakes with width-to-length ratios averaging 0.58 throughout the East Coast and Nebraska.

The impact hypothesis, on the other hand, can explain all aspects of the morphology of the Carolina Bays, including the overlapping bays, but it must also clarify why the geologic dates are inapplicable. From the point of view of the impact hypothesis, even though the OSL dates are very accurate for the terrain on which the bays are found, they do not corresponds to the time of formation of the bays. The answer to this conundrum may be found in the mechanism by which impacts create conical cavities in liquefied soil, and how viscous relaxation modifies the impact cavities.

As illustrated earlier, impacts on liquefied soil create a conical cavity without turbulent mixing, and viscous relaxation restores the stratigraphy by lateral centripetal flow. Only the surface layer of the conical crater is exposed to light, so during the impact and the subsequent modification, the subsoil remains unexposed to light. After the bay reaches its final configuration, the bay has the same underlying physical characteristics as before the impact. The material immediately below the bay surface has not been exposed to light throughout the bay formation process and its geologic date will reflect the date of the original target terrain and not the date when the bay was formed. Thus, using OSL on the subsoil can determine the age of the terrain, but it cannot provide the date of the bay formation event because the subsoil of the bays was not exposed to light by the ballistic impact.

OSL is supposed to measure the time elapsed since buried mineral crystals were exposed to sunlight. The stratigraphic restoration by viscous relaxation leaves a subsurface unexposed to sunlight. Consequently, testing the subsoil of the Carolina Bays will give the geologic date of the target terrain, but it will not correspond to the date when the bays were formed. If all the target material had been mixed and exposed to sunlight at the time that the bays were created, OSL might be an adequate method for testing the age of the bays, but turbulent mixing does not happen for plastic deformations.

It is also necessary to consider that the extraterrestrial impact could have occurred at night in North America. In this case, there would not have been any sunlight to reset the luminescence signal when the Carolina Bays were formed and OSL would not be able to produce any useful results. Similarly, dark clouds and ejecta produced by an extraterrestrial impact during the daytime could have blocked the light of the Sun and prevented the exposure of the terrain to sunlight, which would also invalidate OSL dating.

Figure 79 shows the LiDAR image of Big Bay (Lat. 33.78645, Lon. -80.46785) investigated by Brooks. Is it really true that sand moved across Big Bay about 74,000 years ago and the bay was resurfaced subsequently 33,000 to 29,000 years ago? If the Carolina Bays were made by glacier ice impacts 12,900 years ago, what is the meaning of these dates?

From an impact perspective, we can say that the small bay with a width-to-length ratio of 0.613 was emplaced within seconds before Big Bay with a width-to-length ratio of 0.633. The small bay underwent a plastic deformation in its northwest edge, probably because of the later impact.

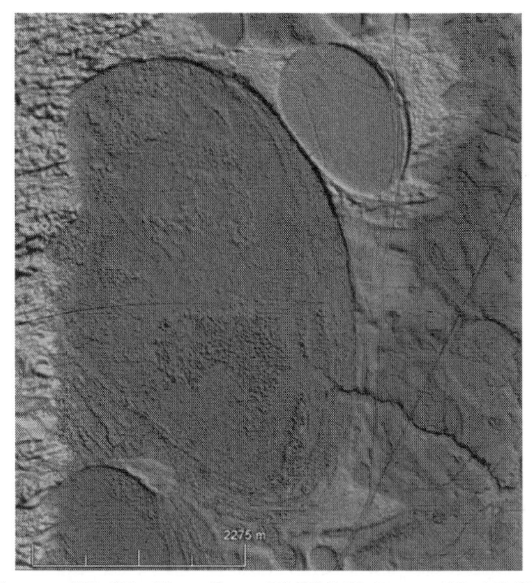

Figure 79. Big Bay (Lat. 33.78645, Lon. -80.46785)

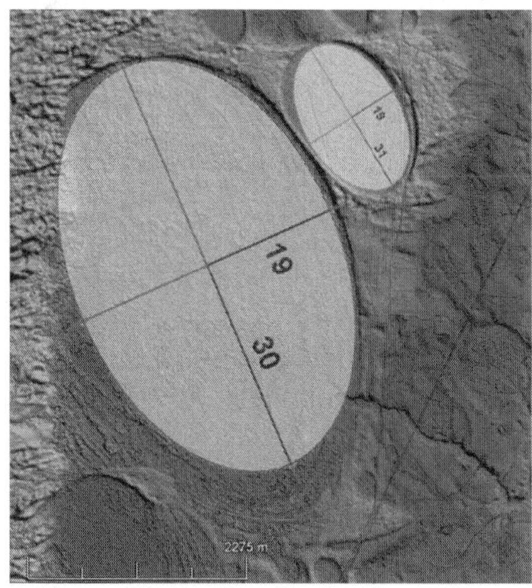

Figure 80. Big Bay fitted with an ellipse

Figure 80 shows drifting soil overlaying the elliptical shape on the west side of Big Bay; the drifting soil is deforming the bay instead of making it more elliptical. A small drainage stream has eroded the rim on the southeast side of Big Bay. Wind and water generally destroy the perfect elliptical shapes of the bays instead of creating them.

The Korjik-Grapes Experiment

To get comments from professionals, I presented this hypothesis in CosmoQuest.com, a forum for geologists and astronomers. I was interested in organizing a citizen-mapping project for the Carolina Bays similar to what is being done for the craters of the Moon, Vesta and Mercury. The title of my first note was "Stop ignoring the Carolina Bays", which became the heading of the thread (Zamora 2013). My assertion that the subsoil of an impact on a viscous surface remains unexposed to light throughout the modification phase was challenged by two CosmoQuest members with IDs *korjik* and *grapes*. In response, I designed the Korjik-Grapes Experiment to demonstrate that only the surface of the conical cavity is exposed to light during the excavation and modification stages of an impact on a viscous surface.

I used a target of equal parts sand and pottery clay with enough water to give a consistency of mortar. The surface was lightly sprinkled with colored green sand to get better contrast. For expediency, I used a marble fired by a slingshot to create the conical cavity instead of an ice ball as I had done in other experiments.

Figure 81. Initial Conical Cavity

Next, I sprinkled red sand on the conical cavity to indicate the surface area exposed to sunlight by the impact. My purpose was to see if any uncolored medium would be exposed by the viscous relaxation process.

Figure 82. Red sand represents surface exposed to sunlight

The photographs document the transformation of the conical cavity into a shallow depression during the modification stage. Throughout the modification stage, only the new surface of the conical cavity is exposed to light but no additional soil is exposed to light. This provides experimental confirmation that the subsurface would retain the same geologic date characteristics as before the impact.

Figure 83. No new surface is exposed to light during remodeling. The geologic date of the subsurface would be the same as before the impact.

The feedback that I got was "OK, so now show that the experiment correctly scales up large scale", and an additional admonition to show exactly how the prior dating method was wrong, and what that would mean to the age of the sites involved.

Need for an alternate method of dating

OSL cannot provide the date of formation of the Carolina Bays because the plastic deformations that created the bays did not expose all of the subsoil to light. The impacts on viscous ground did not free all the trapped electrons in the mineral crystals to reset the OSL clock. Therefore, the dates obtained from the subsoil of the Carolina Bays only reflect the age of the terrain, but not the date of the extraterrestrial impact event that created the bays.

Since the geologic dates of the terrain cannot provide the date of the impact event, it will be necessary to rely on other methods of dating. An impact event of the magnitude indicated by the Carolina Bays must have had a tremendous effect on the environment. The most reliable clues for dating the formation of the Carolina Bays will probably be found in the ecological and archeological record. Events like the onset of the Younger Dryas cooling period, the extinction of the megafauna and the disappearance of the Clovis culture may be the best indicators of the time when the bays were created.

There will be great reluctance to discard all those dates that were so laboriously collected, but eventually, it will be necessary to dissociate the dates of the terrain from the date of the extraterrestrial event.

* * *

ESTIMATE OF THE WATER EJECTED

An extraterrestrial impact on a glacier would have ejected water in its three phases: ice, water and steam. Our discussion thus far has focused on the ice. Can we find out anything about the water? We can estimate the amount of water produced by the extraterrestrial impact by assuming that the duration of the Younger Dryas cooling event was related to the quantity of water ejected above the atmosphere.

Thermodynamics of Liquid Ejecta

An extraterrestrial impact on the Laurentide ice sheet would have produced great quantities of ice, water and steam. Earlier, we calculated that approximately 1.5×10^{12} cubic meters of ice were ejected by the impact on the Laurentide ice sheet; that is enough ice to cover half of the United States to a depth of half a meter. Some knowledge of thermodynamics is needed to find out what happened to the water.

The following phase diagram for water (**Figure 84**) shows that below a pressure of 0.006 atmospheres (611.7 Pascals) water cannot exist in the liquid state. In the vacuum of space, water can only be gaseous or solid. Any liquid water ejected into space will boil, and the evaporation will cool some of the remaining water below the freezing point.

Above 35 kilometers from the Earth's surface, the atmospheric pressure is below the triple point of water and water cannot exist in the liquid state. Any water ejected above the atmosphere or carried along by the ejected ice chunks would have boiled vigorously and produced clouds of ice crystals.

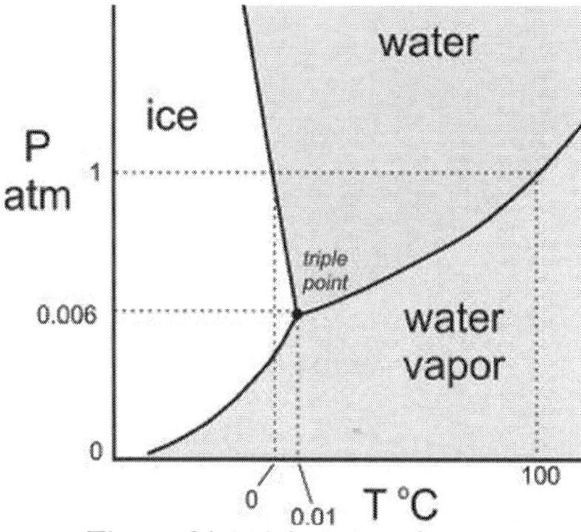

Figure 84. Triple Point of Water

Figure 85 summarizes the effect of energy on the temperature of water in its solid, liquid, and gaseous phases. The heat for melting or freezing water, also called the Heat of Fusion, is 80 calories per gram at 0°C. It is necessary to remove 80 calories of energy to freeze one gram of water at 0°C. The heat for boiling or condensing water at 100°C, called the Heat of Vaporization, is 539 calories per gram.

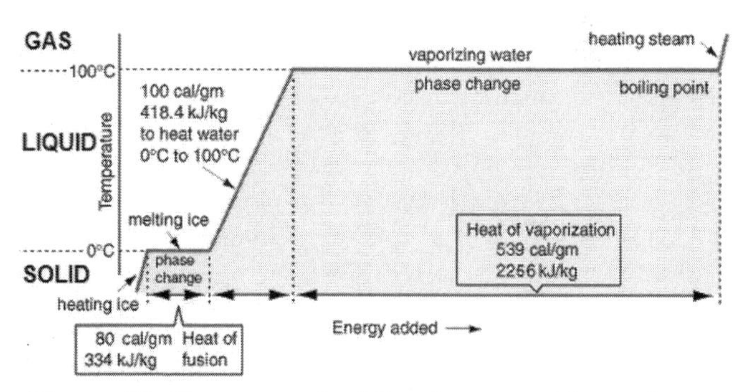

Figure 85. Energy required for phase transitions of water

The water crystallization phenomenon is well known to astronauts when they vent their urine into space. The following is part of a 1976 transcript of astronaut Russel Schweickart talking to Peter Warshall about waste disposal in space:

Schweickart: *Well, actually, in Skylab we did something similar to that. But on Apollo the urine then would go outside, and you'd have to heat the nozzle because, of course, it instantly flashes into ice crystals. And, in fact, I told Stewart this, the most beautiful sight in orbit, or one of the most beautiful sights, is a urine dump at sunset, because as the stuff comes out and as it hits the exit nozzle it instantly flashes into ten million little ice crystals which go out almost in a hemisphere, because, you know, you're exiting into essentially a perfect vacuum, and so the stuff goes in every direction, and all radially out from the spacecraft at relatively high velocity. It's surprising, and it's an incredible stream of ... just a spray of sparklers almost. It's really a spectacular sight. At any rate that's the urine system on Apollo.*

For every 100 grams of water in the vacuum of space, the evaporation of 13 to 25 grams is enough to turn the remaining water into ice, depending on the water's initial temperature. For water ejected at 100°C, it is first necessary to reduce the temperature to 0°C. This requires 100 calories per gram. In addition, 80 calories are required to freeze one gram of water. Thus, we need to remove 180 calories to freeze 1 gram of water at 100°C. For 100 grams of water, we can set up the following equation, where $(100-X)180$ is the number of calories required to freeze $100-X$ grams of water at 100°C, and $539X$ is the number of calories removed by the evaporation of X grams of water.

$$(100 - X)\ 180 = 539\ X$$

$$X = 18000/719 = 25.0 \text{ grams}$$

If 100 grams of water at 100°C are ejected into space, 25 grams will evaporate and 75.0 grams will turn into ice. This means that at least 75 percent of the water ejected above the atmosphere would have formed ice crystals. Using a similar calculation, if 100 grams water at 0°C are ejected into space, 12.9 grams will evaporate and 87.1 grams will turn into ice.

With the additional propulsion provided by evaporation, some ice crystals could have been sent into low Earth orbit where they would have blocked the light of the Sun for many years. This is a reasonable scenario for the onset of a cooling event, such as the Younger Dryas stadial.

The Younger Dryas stadial

As mentioned earlier, the Younger Dryas stadial was a period of cold climatic conditions that started abruptly 12,900 years ago and lasted for 1300 years. Although the calculation of the amount of water ejected by an extraterrestrial impact during the late Pleistocene may appear impossible, some estimates can be made by assuming that the Younger Dryas stadial was caused by blockage of the light of the Sun by ice particles that were formed when water was ejected above the atmosphere.

The amount of ice particles in low Earth orbit can be estimated by the rate of sublimation of ice and the duration of the cold event. Once the amount of ice particles is known, the amount of water ejected and the energy to melt the glacier ice can be calculated. The size of the extraterrestrial object derived from these calculations can then be used to constrain the results to a reasonable estimate.

The standard equation for the evaporation or sublimation rate from a planar surface of pure water or ice in a vacuum is given by S_0 (Andreas 2007), where S_0 is a mass flux; its units are $\mathrm{kg\,m^{-2}\,s^{-1}}$. Also, $e_{sat,i}(T)$ is the saturation vapor pressure (in pascals) for a planar ice surface at temperature T (in kelvins), M_w is the molecular weight of water ($18.015\times10^{-3}\ \mathrm{kg\,mol^{-1}}$), and R is the universal gas constant ($8.31447\ \mathrm{J\,mol^{-1}\,K^{-1}}$).

$$S_0 = e_{sat,i}(T)\left(\frac{M_w}{2\pi RT}\right)$$

The equation for $e_{sat,i}(T)$ provided by Murphy and Koop (2005) for T in [110, 273.15 K] calculates $e_{sat,i}(T)$ in Pa, and T must be in kelvins.

$$e_{sat,i}(T) = \exp(9.550426 - 5723.265/T + 3.53068\ln(T) - 0.00728332T)$$

Assuming that the ice crystals had a surface area equivalent to the surface of a sphere 100 kilometers above the Earth's surface, and given that the Earth's radius is 6371 km, the kilograms of ice crystals can be obtained by multiplying the sublimation rate S_0 times the area and the duration of the Younger Dryas stadial (1300 years). The amount of water ejected can be based on the thermodynamics for converting water into ice; approximately 80% of the water turns into ice crystals. The estimated amount of water depends on the exposed area and the time elapsed, but the temperature has the biggest influence on the result. The following graph shows the amount of water estimated as a function of temperature in kelvins. For reference, 0° Celsius is equal to 273.15°K. These are very cold temperatures.

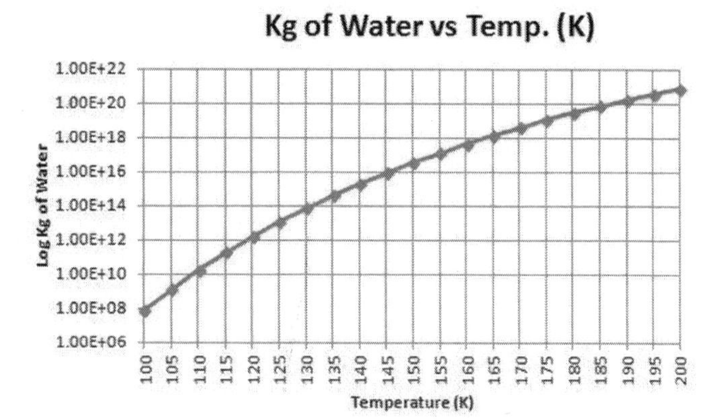

Figure 86. Amount of water based on temperature

The estimates of the amount of water ejected can be used to calculate the energy required to melt enough glacier ice to create the water. A velocity of approximately 3.5 km/sec is required to launch the water into low Earth orbit (~150 km). The energy required to launch the water can be obtained from the formula for kinetic energy $K_e=1/2(mv^2)$. The energy to produce and launch the water, added to the energy calculated to launch the ice boulders (6.35×10^{21} Joules) would be approximately the energy of the extraterrestrial impact.

The following table shows the total energy in Joules required to launch the water and the ice. The total energy is correlated with the size of an asteroid traveling at 17 km/sec. The size of the asteroid increases very rapidly for quantities of water exceeding 2×10^{15} kg. It is unlikely that an asteroid exceeding 4 km in diameter could have been responsible for the creation of the Carolina Bays because such a large asteroid would have left a visible crater on the surface of the Earth even if the target area had been covered by an ice sheet with a thickness of one kilometer. A more fragile icy comet with a diameter of 3 km traveling at 50 km/sec would have had the same energy but would have been less likely to leave a trace of the collision under such a thick ice sheet.

T (K)	Kg of water	Tot. energy	asteroid diameter (km)
100	7.32E+07	6.35E+21	3.04
105	1.25E+09	6.35E+21	3.04
110	1.65E+10	6.35E+21	3.04
115	1.75E+11	6.35E+21	3.04
120	1.53E+12	6.36E+21	3.04
125	1.12E+13	6.42E+21	3.05
130	7.10E+13	6.81E+21	3.11
135	3.92E+14	8.88E+21	3.39
140	1.92E+15	1.87E+22	4.35
145	8.42E+15	6.07E+22	6.44
150	3.35E+16	2.23E+23	9.94
155	1.22E+17	7.96E+23	15.19
160	4.12E+17	2.66E+24	22.73

Figure 87. Estimate of asteroid diameter based on temperature of sublimation

In modern times, the temperature at an altitude of 100 kilometers is approximately 200°K, and it increases with altitude into the thermosphere. The table above indicates that reasonable numbers for the asteroid diameter can only be obtained if the temperature at such altitude could have been 60 degrees lower during the Younger Dryas stadial than today. This may have been possible, considering that the Earth would have had increased albedo from the orbiting ice crystal clouds that blocked the light of the Sun.

* * *

CONCLUSION

The Carolina Bays and the Nebraska Rainwater Basins are remarkable geologic features because of their truly elliptical shape. Such mathematical perfection in their shape and in their radial orientation toward the Great Lakes is indicative of the mechanism that created them. The conic sections encode the story of their origin. In general, the Carolina Bays that are not elliptical were modified by terrestrial processes after their formation or occur in terrain that did not permit the formation of oblique conical cavities.

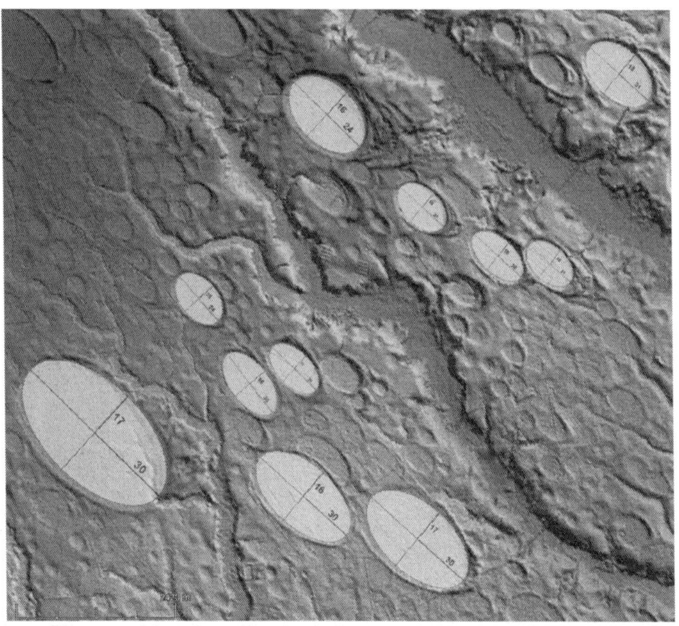

Figure 88. The elliptical shape of the Carolina Bays indicates that they originated as oblique conical cavities that were remodeled into shallow depressions by geologic processes. (Lat. 34.850, Lon. -79.205)

The Glacier Ice Impact Hypothesis proposes a series of mechanisms that could have produced the Carolina Bays by secondary impacts of glacier ice ejected from a primary extraterrestrial impact on the Laurentide ice sheet. The hypothesis has been supplemented with experimental models demonstrating that oblique impacts on viscous surfaces can reproducibly create conical cavities that are remodeled into shallow elliptical depressions by viscous relaxation. This makes it possible to model the Carolina Bays as conic sections whose width-to-length ratio can be explained by the angle of impact. The experiments with adjacent conical impacts also demonstrate a mechanism for creating overlapping bays.

Mathematical analysis based on the number of bays and yield-scaling laws have provided insight about the extraterrestrial object responsible for the creation of the Carolina Bays, which includes an estimate of the energy of the impact and the amount of ice ejected. The application of principles of thermodynamics has provided a plausible cause for the onset of the Younger Dryas cooling event following the impact on the Laurentide ice sheet.

Although the exact location of the extraterrestrial impact remains unknown, the impact hypothesis of the Carolina Bays can be confirmed by finding inverted stratigraphy in the overturned flaps of the bays and by finding stones that might have been carried by the glacier ice chunks. The acceptance of the impact origin of the Carolina Bays will have wide repercussions in archeology, geology and astronomy.

A higher standard for terrestrial hypotheses

Google Earth and LiDAR data have enabled new ways of exploring the geological features on the surface of the Earth. The geometric survey and the application of statistical methods to characterize the Carolina Bays and

Nebraska Rainwater Basins establishes more stringent criteria for the terrestrial hypotheses of bay creation such as substrate dissolution, wind, ice, marine waves and currents that reduce the volume of karst-like depressions which are later modified by wind or ice-push processes. It should no longer be acceptable to propose such hypotheses without at least an accompanying computational model that explains how the conic sections are formed and how the specific width-to-length ratios are achieved and maintained for thousands of years.

The question of the dates

The significance of the dates will continue to be a topic of great controversy. Many scientists have invested entire careers using these dates to show how wind and water created the Carolina Bays. The diversity of the dates has been a barrier to the acceptance of impact hypotheses for the creation of the Carolina Bays. Once geologists and astronomers are convinced that ballistic impacts on viscous surfaces do not alter the date of the subsoil, there will be greater willingness to dissociate the dates of the terrain from the date of the impact event. Improved dating methods will probably emerge from this tumultuous change.

Changes in history and archeology

The density of the Carolina Bays indicates that the saturation bombing from glacier ice boulders could have been the major cause of the mass extinction of the North American megafauna and the Clovis people. It will be necessary to modify the hypothesis that the new human inhabitants of North America exterminated the megafauna or that the Younger Dryas cooling event killed the megafauna. Understanding how the formation of the Carolina Bays altered the surface of the East Coast may help to guide archeological excavations.

Changes in geology and astronomy

Astronomers will no longer be able to say that extraterrestrial impacts have never killed humans. The whole Clovis culture disappeared as a result of an impact. The probability of a killer asteroid will need to be revised, and it will be necessary to recognize that the Younger Dryas cooling event may have been the result of an extraterrestrial impact. Geologists will need to recognize that conic sections in geological formations may indicate the sites of impact events, not only in Carolina Bays and in Nebraska Rainwater Basins , but also in the ring systems of the Aïr Mountains in Niger, which may be the site of a meteorite cluster impact (Zamora 2012).

Future Work

Where was the ET impact?

Determining the location of the extraterrestrial impact is something that will require accurately mapping the bays and extending the major axes to an intersection point. Allowance will have to be made for the Coriolis effect based on the time of flight of the ice projectiles. Variations are to be expected due to mid-air collisions. Research in this area should try to confirm an epicenter in Saginaw Bay in Michigan as calculated by Davias (2015). The mapping of the Carolina Bays is a project where citizen participation may be useful. There are so many bays that one person can only process a small percentage of them.

Let us explore the Carolina Bays

The Carolina Bays are disappearing fast due to erosion, urbanization and farming. To prove the impact hypothesis, it is necessary to search for inverted stratigraphy in the rims of the bays and to dig for rocks that may have been carried by the glacier boulders that made the bays. The exploration of the bays will require identifying the bays that may yield the best information, and then securing the permissions to start digging, since many of the bays are in private property. Any rocks carried by the glacier chunks would be at depths from 100 to 350 meters, depending on the size of the bays, and there is no assurance that every glacier chunk had embedded rocks. If those rocks are found, their geochemistry could help to identify the location of the extraterrestrial impact.

Let us keep our minds open

I feel very lucky to have come across the problem of the Carolina Bays after retiring from a career in chemistry and computer science. I started studying geology and astronomy as a hobby and became interested in impacts on viscous surfaces. By all rights, some professional geologist or astronomer should have developed this hypothesis. Unfortunately, the prevailing attitudes of the impact experts created an antagonistic atmosphere where the study of the Carolina Bays amounted to career suicide with no funding and no support from academia. This provided an opportunity for enthusiasts like me who were not dependent on university or government funding and did not have an advisor who discouraged them from working on the problem or had the power to disapprove a dissertation. There were sneers and quick dismissals when I tried to discuss the topic with professionals. My proposal for study of the Carolina Bays in CosmoQuest.com was assigned to the "against the mainstream" category, almost as if the Carolina Bays were mud puddles unworthy of study. There was plenty of heckling from some participants, and others thought that the ellipses that I had drawn did not fit accurately enough for me to claim that they matched the Carolina Bays "exactly". The useful comments that were made in the forum have been included here.

I am particularly thankful to Michael Davias who spent much time, effort and his own money setting up a LiDAR database for the Carolina Bays and integrated it with Google Earth (Davias 2011). Singlehandedly, he has developed and made accessible to the public some of the best tools for research of the Carolina Bays.

* * *

REFERENCES

Alvarez, L. W.; Alvarez, W.; Asaro, F.; Michel, H. V. (1980). "Extraterrestrial Cause for the Cretaceous-Tertiary Extinction: Experiment and Theory", *Science* **208** (4448): 1095–1108. doi:10.1126/science.208.4448.1095. JSTOR 1683699. PMID 17783054.

Anderson, David G., D. Shane Miller, Derek T. Anderson, Stephen J. Yerka, J. Christopher Gillam, Erik N. Johanson, and Ashley Smallwood 2009 Paleoindians in North America: Evidence from PIDBA (Paleoindian Database of the Americas. Poster presented at the Annual Meeting of the Society for American Archaeology, Atlanta, Georgia, 24 April 2009.

Andreas, E. L, 2007: New estimates for the sublimation rate for ice on the Moon. *Icarus*, 186, 24-30.

Broecker, Wallace S. (2006). "Was the Younger Dryas Triggered by a Flood?". Science 312 (5777): 1146–1148. doi:10.1126/science.1123253, PMID 16728622

Brooks, M. J., B. E, Taylor, and A. H. Ivester, 2010. Carolina bays: time capsules of culture and climate change. *Southeastern Archaeology*. vol. 29, pp. 146–163.

Davias, M., Carolina Bay LiDAR Imagery Viewer http://cintos.org/SaginawManifold/GoogleEarth/LiDAR_Viewer/index.html

Davias, M.; Gilbride, J.L., 2010, Correlating an Impact Structure with the Carolina Bays, GSA Denver Annual Meeting (31 October - 3 November 2010), Paper No. 116-13

Davias, M.; Gilbride, J.L., LiDAR Digital Elevation Maps Employed in Carolina Bay Survey, GSA meeting in Minneapolis, Minnesota (12 October 2011)

Davias, M.; Harris, T., A Tale of two craters: Coriolis-aware trajectory analysis correlates two pleistocene impact strewn fields and gives Michigan a thumb, Geological Society of America, North-Central Section - 49th Annual Meeting (19-20 May 2015)

Dyke, A.S., et al., The Laurentide and Innuitian ice sheets during the Last Glacial Maximum, *Quaternary Science Reviews*, 21 (2002) 9–31

Eimers J.L.; Terziotti, S.; Giorgino, M.; (2001), Estimated Depth to Water, North Carolina, Open File Report 01-487 http://nc.water.usgs.gov/reports/ofr01487/

Eyton, J.R; Judith I. Parkhurst A Re-Evaluation Of The Extraterrestrial Origin Of The Carolina Bays, 1975 http://abob.libs.uga.edu/bobk/cbayint.html

Firestone, Richard; West, Allen; Warwick-Smith, Simon (4 June 2006). The Cycle of Cosmic Catastrophes: How a Stone-Age Comet Changed the Course of World Culture. Bear & Company. p. 392. ISBN 1591430615.

Firestone, R.B., West, A., Kennett, J.P., Becker, L., Bunch, T.E., Revay, Z.S., Schultz, P.H., Belgya, T., Kennett, D.J., Erlandson, J.M., Dickenson, O.J., Goodyear, A.C., Harris, R.S., Howard, G.A., Kloosterman, J.B., Lechler, P., Mayewski, P.A., Montgomery, J., Poreda, R., Darrah, T., Que Hee, S.S., Smith, A.R., Stich, A., Topping, W., Wittke, J.H., Wolbach, W.S., 2007. Evidence for an extraterrestrial impact 12,900 years ago that contributed to the megafaunal extinctions and the Younger Dryas cooling.

Proceedings of the National Academy of Sciences 104, 16016–16021.

Firestone, R. B., The Case for the Younger Dryas Extraterrestrial Impact Event: Mammoth, Megafauna, and Clovis Extinction, 12,900 Years Ago. *Journal of Cosmology*, 2009, Vol 2, pages 256-285.

French B. M. (1998) Traces of Catastrophe: A Handbook of Shock-Metamorphic Effects in Terrestrial Meteorite Impact Structures. LPI Contribution No. 954, Lunar and Planetary Institute, Houston.

Gamble, E.E., Daniels, R.B., and Wheeler, W.H. (1977), Primary and Secondary Rims of Carolina Bays. *Southeastern Geology*, **18**, 199–212

Garvin, J. B., et al., "Linne: Simple Lunar Mare crater geometry from LRO observations", 42nd Lunar and Planetary Science Conference (2011)

Gault, D. E.; Wedekind, J. A., Experimental studies of oblique impact, Lunar and Planetary Science Conference, 9th, Houston, Tex., March 13-17, 1978, Proceedings. Volume 3. (A79-39253 16-91) New York, Pergamon Press, Inc., 1978, p. 3843-3875.

Google Earth. https://www.google.com/earth/

Greeley, R.; Fink, J.; Snyder, D. B.; Gault, D. E.; Guest, J. E.; Schultz, P. H., Impact cratering in viscous targets - Laboratory experiments, Lunar and Planetary Science Conference, 11th, Houston, TX, March 17-21, 1980, Proceedings. Volume 3. (A82-22351 09-91) New York, Pergamon Press, 1980, p. 2075-2097.

Hildebrand, Alan R.; Penfield, Glen T.; Kring, David A.; Pilkington, Mark; Zanoguera, Antonio Camargo; Jacobsen, Stein B.; Boynton, William V. (September 1991). "Chicxulub Crater; a possible Cretaceous/Tertiary boundary impact crater on the Yucatan Peninsula, Mexico". *Geology* **19** (9): 867–871.

Israde-Alcántara, I. et al., Evidence from Central Mexico supporting the Younger Dryas extraterrestrial impact hypothesis, *PNAS*, 03/2012; 109(13):E738-47.

Johnson, Douglas, The Origin of the Carolina Bays, 1942 Columbia University Press

Kennett DJ, Kennett JP, West A et al. (January 2009). "Nanodiamonds in the Younger Dryas boundary sediment layer". *Science* **323** (5910): 94,

LeCompte, Malcolm A.; Albert C. Goodyear, Mark N. Demitroff, Dale Batchelor, Edward K. Vogel, Charles Mooney, Barrett N. Rock, and Alfred W. Seidel, Independent evaluation of conflicting microspherule results from different investigations of the Younger Dryas impact hypothesis, *PNAS* 2012 : 1208603109v1-10.

Maxwell, D.E., 1977, Simple Z model of cratering, ejection, and the overturned flap, Impact and Explosion Cratering, Pergamon Press (New York), p. 1003-1008.

May, James H., and Andrew G. Warne, Hydrogeologic and geochemical factors required for the development of Carolina Bays along the Atlantic and Gulf of Mexico, coastal plain, USA, Environmental and Engineering Geoscience, August 1999 v. 5 no. 3 p. 261-270

Melosh, H.J., 1982, A schematic model of crater modification by gravity, *Journal of Geophysical Research*: Solid Earth, **87**:371-380,

Melosh, H.J., 1989, "Impact Cratering: A Geologic Process", Oxford University Press

Melosh, H. J.; Pierazzo, E., 1997, Impact vapor plume expansion with realistic geometry and equation of state, Conference Paper, 28th Annual Lunar and Planetary Science Conference, p. 935.

Melosh, H.J; Beyer, R.A., 1999, Computing Projectile Size from Crater Diameter.
http://www.lpl.arizona.edu/tekton/crater_p.html

Melosh, H.J., 2011, Planetary Surface Processes, Cambridge University Press

Melton, F. A., and Schriever, W. 1933. "The Carolina 'Bays' - Are They Meteorite Scars?" *Journal of Geology*, Vol. **41**, pp. 52-66.

Murphy, D.M., Koop, T., 2005. Review of the vapour pressures of ice and supercooled water for atmospheric applications. *Quart. J. R. Meteor. Soc.* 131, 1539–1565.

Petaev, Michail I.; Shichun Huang, Stein B. Jacobsen, Alan Zindler, Large Pt anomaly in the Greenland ice core points to a cataclysm at the onset of Younger Dryas, PNAS July 22, 2013, doi: 10.1073/pnas.1303924110

Pinter, Nicholas; Andrew C. Scott; Tyrone L. Daulton; Andrew Podoll; Christian Koeberl; R. Scott Anderson; Scott E. Ishman; The Younger Dryas impact hypothesis: A requiem, *Earth-Science Reviews*, Volume 106, Issues 3–4,

June 2011, Pages 247–264.

Preston, C.D.; Brown, C.Q., 1964. Geologic Section along a Carolina Bay, Sumter County, S.C., *Southeastern Geology*, vol. 6, pp. 21-29.

Prouty, W. F., 1952. Carolina Bays and their Origin, *Bulletin, Geological Society of America*, vol. 63, pp. 167-224.

Raisz, E., 1934. Rounded Lakes and Lagoons of the Coastal Plains of Massachusetts. *The Journal of Geology* 2:839-848.

Ross, Thomas E., 1987, A Comprehensive Bibliography of the Carolina Bays literature, *The Journal of the Elisha Mitchell Scientific Society*, 103(1) 1987, pp. 28-42.

Schulson, Erland M.; The Structure and Mechanical Behavior of Ice, *Journal of The Minerals, Metals and Materials Society JOM*, **51** (2) (1999), pp. 21-27.

Schultz, P.H. (2009) in "Last Extinction: Megabeasts' Sudden Death", NOVA, written and produced by Doug Hamilton, Public Broadcasting System.

Schweickart, Russel (Astronaut) talking to Peter Warshall about waste disposal in space. "Watershed Issue" (Winter 76-77) of The CQ.
http://settlement.arc.nasa.gov/CoEvolutionBook/SPACE.HTML

Shoemaker, E. M., 1960, Penetration mechanics of high velocity meteorites, illustrated by Meteor Crater, Arizona International Geological Congress, 21st, Copenhagen, pt. 8, p. 418434.

Thom, B. G., 1970. Carolina Bays in Horry and Marion Counties, South Carolina, *Bulletin, Geological Society of America*, vol. 81, pp. 783-814.

Wittke, James H; et al., (2013-05-20). "Evidence for deposition of 10 million tonnes of impact spherules across four continents 12,800 y ago", Proceedings of the National Academy of Sciences, *PNAS* June 4, 2013 vol. 110 no. 23 E2088-E2097, doi: 10.1073/pnas.1301760110

Zamora, Antonio, 2012, Meteorite Cluster Impacts, Kindle eBook (ISBN:978-0-9836523-2-8, January 11, 2012)

Zamora, Antonio, Interpreting Carolina Bays as Glacier Ice impacts, June 28, 2013, http://www.scientificpsychic.com/etc/carolina-bays/carolina-bays.html

Zamora, Antonio (ID: citpeks), Stop ignoring the Carolina Bays, CosmoQuest.org forum, (Nov. 2013) http://cosmoquest.org/forum/showthread.php?153071-Stop-ignoring-the-Carolina-Bays

Zamora, Antonio, Killer Comet: What the Carolina Bays tell us, Kindle eBook (ISBN: 978-0-9836523-6-6, January 2, 2014), Paperback Edition (ISBN: 978-0-9836523-7-3, January 1, 2014)

Zanner, William and Kuzila, Mark S., Nebraska's Carolina Bays, (GSA Annual Meeting, 2001)

* * *

ABOUT THE AUTHOR

Antonio Zamora has a multidisciplinary background in chemistry, computer science and computational linguistics. He studied chemistry at the University of Texas, and Computer and Information Science at Ohio State University. During his service in the U.S. Army, Mr. Zamora studied medical technology and worked in hematology at the Brooke Army Medical Center. Mr. Zamora worked for many years as an editor and researcher at Chemical Abstracts Service developing chemical information applications. He also worked as a senior programmer at IBM on spelling checkers and novel multilingual information retrieval tools. He was the author of 13 patents. After his retirement from IBM, Mr. Zamora established Zamora Consulting, LLC and worked as a consultant for the American Chemical Society, the National Library of Medicine, and the Department of Energy to support semantic enhancements for search engines. Mr. Zamora has been interested in astronomy since childhood when his father helped him build a refracting telescope. During retirement, Mr. Zamora has completed massive open online courses in astronomy, geology and paleobiology. He regularly attends the seminars of the Department of Terrestrial Magnetism at the Carnegie Institution of Washington.

Printed in Great Britain
by Amazon